Dammit Peckerhead:

Tales from the Homestead

By N.A. Broadley

Angry Eagle Publishing
https://angryeaglepublishing.com

Printed in the United States of America
First Printing, 2020
ISBN: 978-1732621282

No techniques are recommended without proper instruction or safety measures and training. The author nor publisher assumes no liability for any action presumed from this book.

Cover Design: Piper Leigh Creatives
Editorial and formatting provided through Angry Eagle Publishing.

https://angryeaglepublishing.com

Editor's note:

This is a work of stories compiled from social media posts over a period of time. While every attempt was made to correct grammatical errors, many remain due to the desire to maintain the integrity of the story.

Please enjoy the story for its hilarity and testament to the lifestyle.

Contents

The Homestead

When I first saw the old cabin, my heart skipped a beat. It was perfect, probably for no-one else, but it was for me. I was coming out of a bitter place. A place that left me hopeless and questioning, tired physically, mentally, and emotionally, and worse... soul tired. The cabin was my escape.

It would be my sanctuary from losing a brother to suicide, from battling a deep, soul depression of becoming an empty nester and from losing so much that I had worked so hard to build. I was ready to leave behind all noise, all the clutter, all the sadness and throw myself into solitude where I could sweat and toil, bleed and scream, heal and find my laughter again.

And the cabin, this dilapidated, mice ridden, building of aged logs and cobwebs gave me the path on which I could do this. These stories that I share with you are of my homestead beginning. They are of finding my way again in a huge world where I'd become so lost.

Rough Beginnings

The rooster, aka Peckerhead, began crowing at 3 a.m. I rolled over in bed and sighed tiredly, looking at the alarm clock on the bedside stand. It had been retired from its thirty-year job just this past year. No longer needed. To say I worked one job for thirty years would be laughable. No, I worked many jobs, walked many hallways on tired legs and worse, with a tired and disillusioned heart before I finally gave in and decided to chase my passion.

Homesteading, writing, and living a life of simplicity. So, at fifty-something years old I embarked on this journey and have not looked back. Do I miss my working life? Not a bit! Do I love my homesteading life? With every ounce of my being, every cell in my body and every contented whisper of my soul. And as I lay in bed, listening to Peckerhead greet the morning as only roosters do, I smiled. Tired or not, the animals needed tending, the chores needed starting. And as every day had begun over the past year, I crawled out of the warmth of the blankets and greeted the day with my own unique call.

I will be sharing stories with you about my day to day life here on the homestead although many of these stories tell of the lighter side of this life, the more humorous of daily events, I will be honest with you, it hasn't been all sunshine and roses. In fact, it is hard. It is bloody and painful, sweaty and dirty, heartbreaking and defeating. Yes, there have been times when I've questioned my sanity in pursuing this lifestyle.

But it is also rewarding. It fills my cup with love, laughter, fulfillment, and simplicity.

In August 2016 I traded in my nursing scrubs for ratty jeans and my high heels for work boots. My paycheck for a dream. Some would look at my lifestyle and think poor. But poor, like beauty, is in the eye of the beholder. After reading my story, I will let you decide.

I stood in front of the log cabin that my husband and I had just purchased. We paid a slight song and dance for this little cabin and the piece of dirt it stood upon. When I first looked at the cabin in June, my heart stopped. It was the perfect image of what I had carried in my heart and dreams for so long. The first step through the door and I knew I had found my home.

The cabin had been empty for years, run down and begging for someone to give it some tender loving care. It had no insulation. It had drafts everywhere. It was unfinished inside. A well that was nearly dry and no modern septic system. The only heat source was a rusted old wood stove. Was I just plain crazy? Had I lost my mind? Yes and no. I knew this was where I was meant to land. Here, where I could sweat and work, write and learn, wander and be found. Crazy? Poor? Blessed? A matter of perspective I say.

Sweat, Blood and Toil

I can't flush the toilet. The well is low, and we are in the middle of a summer-long drought. I had to carry the sudsy pan of graying dishwater to the toilet just to flush. I was quickly learning to conserve. A dry well is a nightmare at this early stage. After days of plumbing problems, we now had running water, and we could not afford to waste a drop. Summer sizzling heat drove me to crave a long and cool shower, but I would settle for a sponge bath later in the evening. I was sweating just sitting still. I had hornets congregating a meeting in the upstairs loft. I wondered what they were talking about and I hoped they didn't find their way downstairs. I didn't mind sharing my house with the little critters as long as they stayed in their space and left mine alone.

Sweat trickled down my back as I crawled under the house. It was a four-foot-high space, no underpinning or foundation. Post and beam are holding up the cabin, dirt floor underneath.

Littered with the remains of the last tenant's possessions, skeletons of mice and other small critters, I made my way through the shadowed crawl space. Garbage bags and old boards, debris, and other trash litter the dirt floor. Project number one I had decided was to clean it out and insulate. Otherwise, we might be looking at a very cold house come winter.

Why in the world did these people (the former owners) build a rock wall around the entire house as skirting? It was pretty, yes, but not very practical. Who does that? So, I at fifty plus years old, found myself crawling around in the dirt, in the dark because of that stupid rock wall, hearing the skittering of mice and praying one did not jump on me and get inside my shirt or pants. If that happened, I knew, I would just scream and die. I was covered with dirt, cobwebs and mice poop and I can't help but laugh as I look happily at the new direction of my life. This surprises me, the sound of my laugh. I hadn't laughed in a very long time. This is the stuff dreams are made of. Mice shit and all.

A Reprieve

There is a laughter that dances in my soul. This

homestead, these warm, busy, sunny days where I am surrounded by birch, maple, tall pines, and spruce trees dropping their pinecones and acorns, is just the balm I need. My mind slows down, taking on the peaceful pace around me. Every morning I wake before the sun comes up, move to my front porch with a cup of coffee in hand and welcome the day like I would an old and dear friend. My heart is settled and my search over as I have found what I've needed all along. Just this soul balm. This sweet, sweaty, hard-working lifestyle of this homestead. Herbs: basil, mints, rosemary, catnip all thrive beside the strawberry plants which are flowering soon to grace me with ripe, red berries. Flowers, purple and white, like colored snow dusting the green of the lawn….

Today this is enough, in fact, it is more than enough, and I am awash in gratitude for what I've come home to. Myself and my passion for these two-and-a-half acres….

Critter Junction

Another day always comes with another set of challenges. I have decided that my body hates me this morning and wants a divorce. I can't say as I blame it. I ache in places that haven't screamed at me for quite some time and I move like an eighty-year-old woman with arthritis and bunions, wobbling and limping.

But… ahh… the moon is beautiful. The glow casting over the front yard is a pale, serene light that washes over me. Up before dawn I gently ask my body to cooperate and begin doubting my sanity for taking on such a monumental task as homesteading. I have rocks to move today. I know each one is a potential back-breaking, muscle busting, and evil little bastard. The plan is to remove all the skirting wall rocks and replace them with wooden skirting around the cabin. I try to look at it as a good work out for the day. Hell, I don't have to join a gym; I get plenty of weightlifting exercise right here at home. Home. This word makes me happy.

"Ma, wrap the hook and chain here, and I'll give it a pull," yells my son.

He hands me a chain that weighs more than I do. It is already eighty plus degrees, and I'm sweating. And I stink. I hadn't realized a person could smell this disgusting. But I sure do and it ain't a pretty.

I do what my son asks and warn him to wait till I am out of the way before he gives the chain a yank with his four-wheeler. Not that I don't trust the kid, but hey, we don't want me losing any fingers in this exercise of insanity. Who in the hell would build a rock wall around the perimeter of their house? I just want to scream with frustration. To make it even worse, they cemented the freaking rocks together! So not only am I trying to move heavy rocks, I am also having to swing a sledgehammer to break them apart! I am too old for this poop. But, in my defense, I am insane and following my dream, so this is all okay.

I switched the sledgehammer for a small hammer. Step one, pull out a chunk of the rock wall with the four-wheeler and try not to lose any body parts in the process. Step two break apart the rocks from the cement and try not to lose any body parts in the process. Step three, load the rocks in the wagon and try not to rip or break or tear or wound any muscles in the process.

Isn't this what they do in prison? Those black-and-white striped convicts of the rock gang? Or is it a road gang? I wondered if I should break out in song, something solemn and low. I slam the hammer into another hunk of rock and cement trying to break it apart. Of course, it is my luck that at this moment a mouse decides to seek safety on my lap. His or her expression is one of sheer surprise, my own, if I could see my own face, I know would be one of horror.

I yell jump up and do the dance of terror. My son, wonderful boy that he is, is rolling on the ground laughing. And I am feeling like I have things crawling on me. "That's it! I'm done!"

Between great chortles of laughter, he chokes.

"Ma… it was just a tiny mouse," he snorts.

I scowl at him and growl under my breath. There are two things in this world that have the power to make me scream like a little schoolgirl…. Mice and Spiders.

"We need a break." I said as I stomped off in a fit of temper.

I am sweaty and filthy; my long hair is full of rock dust and spider webs. And my back feels as though it is ready to snap in a thousand different places. Between the spiders and the mice, I imagine the worst. All I want is a cup of herbal tea, a cool shower, and air conditioning. Then I remember I have to conserve water, and hell; I don't have an air conditioner. A cold washcloth would have to do. In defeat, I call my sister and shamelessly begged a favor.

"Can I use your tub?" I ask.

Such is the life up here on the hill.

Visions of a cool bath floated through my mind for the rest of my busy afternoon. Sunburn has turned my arms a laughable bright red. My fingernails, once polished and pretty, are chipped and broken. My hands are dirty and nicked with little cuts from the rocks. I did not mind a bit. I was following my dream of living as a homesteader. Or was it a nightmare? Nope, it was a dream. At the end of the day for the first time in a long time, I swam in the feeling of accomplishment. I worked hard; I sweat, I swore and bled but I did it. I moved one-quarter of that blasted rock wall. I made a mental note to add more bandages to my shopping list. And it was good.

Sigh

The work is good. My sleep is better. And as I greet each morning, I can feel the remnants of the day before in my body. Strength and soreness remind me I am hitting my stride. Stretching and strengthening muscles that haven't had this kind of work out for a very long time. It feels good, a purposeful way to spend my days toiling in the dirt, building and cleaning, fixing and destroying. I felt as though those dark years of losing myself to the chaos that was my life is now making sense to me. I walked today the length and depth of my property. I touched the trees, sank my feet in the swampy mud, picked flowers, and listened to the whispers of the woods. I climbed onto the stone walls and sat in silence absorbing the life unseen but heard as the wild hid from me.

And I adored every moment of exploring, of wondering, and of feeling and breathing.... Oh, so deeply.

Shake, Rattle, and Bite

The previous owners, whom now I had developed a strong disliking for, had graced us with a shed that was dilapidated and collapsed. It was an eyesore announcing its place in my backyard like a goliath sized insult. My mind said I could do it, but my body warned with a little hiss that I was crazy even to try.

So, throwing on the rattiest pair of pants I could find and a tank top, I made my way to the backyard to tackle the colossal beast. It insulted me just being there. I knew it was probably full of critters, ones that would give me nightmares and have me doing the DANCE OF TERROR. I just knew that before this day was over something nasty would be crawling on me.

Grabbing a crowbar, a shovel, a rake and my trusty hammer I glared at that dilapidated mess and confidently walked over to it. I looked down at my fur-girls who were dogging my heels.

"Okay, if mama dies doing this please don't chew on my dead carcass okay?" I muttered.

They gave me the look. Sucking in a deep breath, I eyed the mess for bees, mice, skunks, or any other creature that could be housed in the debris. God help me if a hairless rat-like opossum skittered out after me. I just knew that if that happened dying would be a very real possibility.

I told myself, actually an attempt to encourage myself, that I had this. Easy peasy right? I had watched every episode of Alaska, The Last Frontier; this was nothing compared to what they had gone through on their homestead. Reaching in, hoping I wouldn't grab something that would bite me, I began pulling boards and rotted pieces of wood from the tumbled down shed. Out came trash bags that I just knew were full of mice. Out came old coolers and camp stoves. Out came shattered clay pigeons. Out came old blankets and sleeping bags…. WHO WERE THESE PEOPLE? WHAT KIND OF PERSON LEFT A MESS LIKE THIS FOR THE NEW HOMEOWNER? WHERE WAS THE PRIDE? THE CONSIDERATION?

I felt the pain of the sting before I saw the culprit. Red ants! Crawling up my legs, all over my gloved hands, on my arms. Really? Marching Red ants? Attacking me! I hit the mother lode, yay me.

In my dance to get away, I stepped back onto a rotted board watching in horror as my foot and leg punched through the floor. I envisioned an attack from below, some woodland creature chewing on my ankle. Panic enveloped me as I struggled against all the monsters I just knew lurked beneath the rotted floorboards. It could be anything! That opossum that was giving my wild imagination fits or the toothy, grinning clown from Stephen King's, (It), or a tarantula waiting to inject me with its venom and my husband would come home to find me swollen and purple, choking on my tongue. Okay, yes, I am exaggerating a bit. Drawing a deep breath, I realized I needed to settle down and pull my big girl pants up. I pulled my leg out and examined it for damage. Not bad, a little bit of blood from a scrape and nothing hanging onto it hungrily. I could handle this.

Board by board I tore that shed apart. Wild woman, mad woman, or just maybe… Crazy woman. I was relentless. I was determined. And after hours of working under the punishment of the hot sun, amidst the ants, rusty nails and rotted boards, trash, and unseen creepy crawlies, glancing nervously into the woods around me for that crazy (It) clown, I finally exposed the skeleton of the beast. Proudly I stood back and looked at my piles. One of the piles consisted of boards I would be saving for re-use and the other, a pile that would be loaded and hauled off to the landfill.

"Why don't you call for a dumpster Ma?" my son suggested as he looked at the piles of debris prettying up the back yard.

I hadn't thought of that, but it sounded good to me. A lot less work than having to load the truck and haul it all off myself. So, getting on the phone, I placed the call.

"Hi, could you tell me how much you would charge to deliver a dumpster to my house?" I asked pleasantly.

"Sure. Four hundred dollars for the month."

"What!!! That's outrageous! How do you justify that for a metal box?"

"It is what it is. Do you want the dumpster or not?"

At this point, I was trying to weigh the benefit of having it or not. These people were crazy! That was a car payment for God's sake! That was a lot of money!

"No, I certainly do not! Thank you, but I think I'll keep my money." I replied snottily and hung up the phone.

I proceeded to load the truck with the blasted, red ant mess of the partially torn apart shed. Sweat poured down my back, dust clogged my nose, and those freakin ants kept getting feisty and chewing on me. There wasn't a spot on my body that didn't hurt. And I looked at the mess, looked at the truck, swore and sweat.

"Give it up Ma… get the friggin dumpster," my son pleaded.

"By the time you spend hours loading the truck for the dozen trips it'll take you taking this crap to the dump, you'd of paid for the dumpster two times over."

I grudgingly knew he was right. By the time I put gas in the truck and added in the hours spent loading it, hurting myself, bleeding on it and running back and forth to the town dump, it would probably equal what I would pay just to have a dumpster delivered to fill and have hauled back off when I was ready.

Getting back on the phone, I placed the call. After giving them such a hard time earlier, I humbled myself to their mercy, and I disguised my voice and ordered the dumpster. Lesson learned, pay one way or pay another, choose your level of pain. And it was all pain.

Calendula oil became my best friend on this day as I dabbed a cotton ball soaked in it on every red ant bite. Swearing and sweating the whole time. By the time my husband got home from work I'd moved another wagon load of rocks, tore apart most of the collapsed shed, stepped on a snake with bare feet, chased wasps from my bedroom and been bitten more times than I could count by evil little red ants and did battle in my mind with that crazy clown I just knew lurked in the woods behind the shed.

But happily, and much to my relief, I hadn't run into the hairless, rat like thingy of my imagination. That was a plus in my book. I stared in pissed off defiance at the ugly green dumpster that now sat in all its glory in my front yard.

"So, you went with the dumpster idea, huh?" he said laughing. He knew how much it pissed off my inner cheapskate having it there.

"Yup, and I'll load that thing to the high heavens." I growled as I thought of getting my monies worth from the monstrosity. Oh, that company would pay for charging me so much!

"Wow, you did a lot today," he said as he looked around the yard.

A big pile of rocks stood in the area where next year I planned to have my garden, neat stacks of boards in the backyard and yes, even a neat stack of debris that hadn't yet made it to the dumpster. It all had my DNA attached to it. Blood from the cuts and nicks on my hands. I had battled the beast and conquered it. I had not been eaten by my dogs! I now owned it!

"Honey do you really want to save the boards?" he asked.

I glared at him like a bear protecting its cub, daring him to challenge me. I bled for those boards! I had done battle with monsters that he had no idea were lurking around our house. All for those boards!

"Yes!" I replied calmly.

"We can buy new ones."

"Or…. we can use them and recycle," I argued.

"And what are you planning on using them for?" he asked.

"My goat pen, my chicken coop, and the wood

bins," I replied.

He shook his head.

"You're insane," he said.

I smiled.

Yup, I knew it, he knew it, and it was all good.

"And that is why you love me," I teased.

Standing there with my hair a mess, covered in dirt and dust, sweaty and probably not smelling like sunshine and roses, I realized how good I felt. How happy. My body felt like it had been through the proverbial ringer and I was so exhausted I felt actually dizzy with it, but my heart was happy. I plotted my next move of getting rid of the ants, gleefully, with malice. Those freaking little bastards had to go.

Blood, Sweat and Roars

Early morning, before the sun kisses the sky and darkness is reigning, is the most favorite time of day for me. The eerie call of the coyotes, the stirring of woodland creatures. I realize that on the homestead there is always work to be done. And today would be no different.

After battling the red ants, the imaginary killer clowns, and the opossum I feared was stalking me, I finally got the wrecked wooden shed torn apart. It took me days to pull nails from the existing boards that I was saving for other building projects. Who in their right mind build structures with nails? Any common sense builder knew to use wood screws. And the former tenants, I had decided, didn't work with much common sense when building.

The project staring me square in the eyes was building a woodshed from old boards and collected pallets. Creaky bones, sore muscles, and a call from my inner Rosie the Riveter lifted me from my comfy chair on the porch. I looked down at my fur-girls and once again warned them not to eat me, should I die from this latest task before me.

Dragging my weary ass to the backyard, I began gathering the boards I would need. My plan was simple. Attach one two by four plank to each corner of the pallet. Then brace each board with one smaller board across the top and the bottom forming a box like structure. I had two cords of wood that needed to be stacked and I was becoming very tired of looking at it sitting in a pile in my front yard. Wood piles attracted snakes and mice. Snakes I didn't mind but mice? Well, let's just say I don't like sharing my space with those little critters.

I plotted and planned where the best place would be so that I would have the most convenient access to my wood bin. I decided to place it just on the edge of the front yard, not far from my front porch, on the edge of a thick patch of blackberry bushes and brambles.

I looked at the pile of boards stacked neatly beside the pallets. In my hand I held my husband's wireless drill with what I hoped was the proper screw bit. Board one. With one knee holding the board in place against the pallet, I twisted my body into a pretzel. Not good for someone my age. I shoved my shoulder against the board for more stability, screw in hand and in the other, the drill.

I placed the screw against the board where I

wanted to attach it to the pallet. I placed the drill against the screw and pressed the trigger. To my surprise the drill slips and drives itself into my finger and I tumble into the clustered and evil patch of blackberry brambles. Ouch and ouch. Thorns gleefully dragged themselves across my bare arms and my finger is dripping blood. Okay, so this isn't going as planned.

Soaking up the blood with the tail of my shirt, I tried again. Again, the drill slips and I stab myself but thankfully I don't end up in the briar patch. What came out of my mouth would have made a truck driver blush. I threw my husband's drill across the yard in a fit of temper and my fur-girls ran for cover. No wonder people used a hammer and nails. I would have to re-think this project. If not, I feared I would either end up in emergency getting stitches or possibly losing a finger or two. I wondered if my hubby would still love me if I had no fingers.

Needless to say, I don't give up easily. Should I stand the pallet on end and try to drill the screw down into it rather than kneeling and twisting like a pretzel drilling in sideways? Lifting the heavy pallet, I tipped it up on end and held it vertical with my hip. Placing a board on one corner flatly I held it up with my knee.

Okay, so this has got to work and fast, standing on one leg while my other knee was raised to hold the board in place was not an ideal position for me. Grabbing another screw, fitting it into the drill tip I pressed it into the board. This time it bit in and I drove that screw deep into the board. Albeit slipping the screw head the entire time and rounding it out. Sweating and breathing heavy I quickly screwed in two more screws. Okay, so attaching one board had taken me approximately an hour. Boy, was this going to be a long day.

By the time I got the first wood bin built, I learned several things. Number one, I love my fingers and didn't want to lose any. And with the accomplishment of building this first wood bin I now sported a band aid on each finger and numerous scratches prettying up my arms and face. Number two, I needed to buy another box of Band-Aids. Number three, when you have a temper tantrum do not throw your tools across the yard and scream. What will the neighbors think? And lastly, number four, your body is not meant to twist and contort like a pretzel. You will pay for it later.

All in all, it took me eight long, hot, bloody hours to build one large woodbin to hold the first few cords of wood. It was a rickety eyesore. It was a beastly nemesis that did its best to thwart me at every turn. I cried, I swore, I sweat, and I bled for this monstrous contraption. When the last screw was drilled, I stood tall and defiant, bloody and beat to shit, sweaty and covered with grass and dirt and mashed blackberries. I raised my arm holding the drill high above my head and gave a primal, confident scream. I was a hot mess and of course it was at that exact moment when my neighbor decided to stop by and introduce himself.

I watched the truck pull into the yard, and I froze in embarrassment. One arm held high like Zena the warrior princess, clothed in ratty, torn shorts and mismatched socks, worn work boots and a tank top. Dirty, bloody, sweaty. Should I run and hide? With as much dignity as I could muster, I lowered my arm and smiled. Perhaps he'd heard my screams up on the hill and came down to rescue me.

"Can I help you?" I asked, trying to look

dignified in my messiness.

A man, about the same age as my son, climbed out of the truck. In his hands he had two dozen eggs.

"Welcome to the neighborhood," he said with a smile and handed me the eggs. "These are from my farm up the hill. I hope you like eggs."

I did like eggs; in fact, my daughter had just given me two dozen the day before. Smiling, I took the eggs and wondered just what in the hell I would do with four dozen of the blasted things. Oh, boy, hubby was going to get a lot of egg dishes for dinner over the next week or so. Laughing, I could just imagine the conversation my new neighbor would be having with his wife over dinner that evening. It probably kind of went something like this.

"Honey, I met the new neighbor. You probably shouldn't bring the kids down to meet her. I think she might be a bit on the crazy side."

Such is my life on the hill. There is a reason I am an introvert. And I am certain that my neighbor does think of me as the crazy lady that lives at the bottom of the hill.

Implosion

I glared at my hubby like I was the spawn of Satan ready to drag his bony ass to the confines of hell. I'd had enough, and I wasn't taking any more!

Two days prior I spied a fitting on the water pump leaking. Just little drops. Insignificant, really.

"Honey, the fitting near the water pump is

leaking," I said after he'd gotten home from work.

"You know as well as I do, a little leak if not fixed will turn into a major problem eventually."

Tired from work, he just nodded.

"Well?" I asked.

You see my husband is the world's biggest procrastinator, and it drives me absolutely bat shit crazy at times. To the point really where I feel that the top of my head is going to pop off. Taking the hint, he grabs a big wrench.

"Okay, well maybe it just needs to be tightened."

My water pantry is only a four by four little room, barely big enough for the both of us to squeeze into. But, while he was turning the wrench, trying to tighten the fitting, I needed to hold the pipe, so we didn't end up breaking it. Or worse, wrenching the whole set-up loose. Now me, I would have grabbed a roll of duct tape and just taped around that little sucker, but no, he insisted it was just a loose fitting and tightening it would do the trick. Mind you, these pipes hadn't seen water flow for going on ten years and there was quite a bit of rust and tender spots around the fitting.

"Kay honey hold that tight, so it doesn't move," he said as he placed the wrench on the fitting.

Ohhh… I just knew that of course it wouldn't be as simple as it seemed. Turning the wrench, I heard the fitting squeal and screech.

"One more turn should do it," he muttered as he gave it one last turn.

One more turn. Yup. It sure did it all right. Instead of an insignificant little drip we now had a gusher. Jumping back, I landed on my husband who in turn landed on some shelving in the water pantry, knocking jars of canned foods onto the floor where they exploded in quite the pretty mess. I had peaches, tomato sauce, baked beans, green beans, chili and corn covering the floor and water pouring in like we'd opened the gates of soup hell. It was a raining, soggy, soupy mess. And we both stared in amazement, in shock at what had happened, I glared at my husband. Really? Really? You've got to be kidding me! I was standing ankle deep in soup.

"Hurry turn off the main valve!" my hubby yelled as he hopped, skipped and danced away from the mess.

Reaching over I shut the main valve off essentially shutting the whole house down. It was then that the water pump gave a warning growl, a thump and started to merrily sizzle. My first thought was, 'Oh, my God it going to blow! Run!' But I didn't run, I yelled instead.

"Honey shut the breaker down to the water room," I yelled as I waded through the peaches, tomato sauce, baked beans, green beans, chili and corn.

It took me two hours to clean up the mess of one little dripping fitting. Two hours of sloshing through vegetable soup, mopping it up with towels.

Calling a plumber, I begged shamelessly for him to come out and fix the leak. It took two days for him to show up, two days of bucketing water from the well to the house to do dishes, to cook with, to wash with and to flush the toilet. I felt like a Pioneer woman. Yes, hear me roar… or at this point, whimper.

This was our first fiasco with the plumbing in our cabin, but it sure as hell wouldn't be our last. It wasn't a short week later that I was in the middle of taking my shower, and trust me my quick ten minute showers were like Christmas day to me as the drought and low level well had taken away my long luxurious soaks that I had been used to, when suddenly my steamy hot water turned icy cold.

"Honey, I've no hot water," I yelled as I rinsed quickly.

Our hot water heater had suddenly quit. Shit! The following day we made a trip to the hardware store and purchased a new hot water heater.

"Want me to call the plumber and see if they can come out and hook it up real quick?" I asked hubby as we were driving home.

Not that I wanted to spend the money, but lately everything hubby touched went to shit and I had visions of more flooding in my house if he tried to do it himself.

"No, I got this. I mean really how hard can it be?" he quipped.

Oh, no, really this was not what I wanted to hear. But I also didn't want to challenge his manhood or some other foolish thing.

So, hauling the empty new hot water heater in and struggling to get the old hot water heater out, we managed to get it set into place without throwing any punches, Hubby hauls out his handy dandy wrench and sets to work. I cautiously watch over his shoulder, towels ready in case of an explosion. First aid kit on the kitchen table and 911 on speed dial.

"Okay, turn on the pump breaker and let's get this thing filled up," hubby says.

I hesitate.

"Shouldn't we read the instructions? They gave us a booklet for a reason," I suggested.

"Nah, we just need to turn on the water, fill the tank and turn on the breaker to heat it up," he replies, somewhat proud of his work.

"Okay…." I said and proceeded to turn on the pump.

I heard the sound of precious water filling the empty tank.

"Don't forget the hot water heater breaker honey," hubby reminds me.

I flip the breaker for the hot water heater.

"Kay we should have hot water in two hours," hubby assured me as he looked for leaks and drips in the water lines.

I breathe a sigh of relief when I saw there were none. I went about my business for the next few hours just happy with the fact that soon I would have hot water.

"Honey, the water heaters been running for five hours now and we still aren't getting any hot water," I complained.

What did we do wrong? Did we get a bad water heater?

"What the ****" he sputtered, looking helplessly at the water heater. "I know I hooked it up right. Maybe the breaker is no good. Better shut it down honey," he muttered.

Perplexed we looked at each other. Grabbing the instruction manual, I read it from cover to cover. Yup, I think we did everything right. But hell, I was no plumber, and the instructions didn't make much sense to me.

"I'm calling the plumber," I said.

Eight hundred dollars later and a very happy plumber informs me that unlike gas hot water heaters, electric ones needed to fill completely before you turned the power onto them. Our new hot water heater had burned out the top electric element because we failed to fill the tank completely. He then proceeded to tell me that it was all in the instruction manual that came with the new water heater. I looked at my husband with an expression that should have brought terror to his heart, then I looked at the rifle hanging on the wall above the television.

"I could so fudging shoot you right now!" I said angrily as I glared at him.

To my surprise, he started laughing.

"Then I guess it's a damn good thing you are too short to reach the gun, eh?" he replied between great gusts of laughter.

I stormed off to take a hot shower.

Thus far, in one week, we had paid out four hundred dollars for a new hot water heater and over one thousand dollars to our friendly neighborhood plumber. I think actually we bought him a new truck at this point. But we weren't quite done yet.

My plans for that Friday were to clean the house early and then give myself a much-needed break and have lunch with my sister. I was looking forward to getting away from the wood, the building, the plumbing issues, the mice, the snakes, the red ants, the scary clown like thing hiding in my woods, and the opossum that had yet to appear.

As I was washing the floor in the bathroom, I noticed a small puddle of water near the bathroom sink. Giving it no thought, I mopped it up only to have it reappear a short few minutes later. My heart sank, and I wanted to sit on the bathroom floor and just cry. Opening the vanity doors under the sink, I spied the culprit. A small drip.

"Honey, the bathroom sink pipes are leaking," I said. "I'm calling the plumber." Before I could pick up the phone to dial, my husband had picked up his trusty wrench.

"No, no, no!" I yelled as he headed into the bathroom.

"Let me just try to tighten the fitting. I'm sure it won't be as bad as before," he replied with glee in his eyes.

He loved to play handy man. I had visions of flooding, rotted floorboards and murder. Grabbing every towel I could find, I followed my man into the bathroom.

"I'll just tighten it a bit," he said as he lay under the bathroom sink.

I wasn't surprised as I watched a quiet little drip turn into a geyser that shot toward the ceiling. I completely and insanely lost it! I threw the towels at him and yelled for him to clean up his own mess. I snatched the wrench from where he places it on the floor and ran out the front door with it.

With a scream of frustration, I launched it into the woods then picked up several rocks and threw those too. Marching back into the house I looked at my husband who was now on his hands and knees mopping up several inches of water on the floor and I threatened him.

"You touch one more thing in this house and I swear to God I will murder you when you go to sleep and bury you in the backyard! Oh, and I will also divorce you," I screamed.

Once again, I had no water to the entire house. This lasted five days before my plumber, Mitch, and yes, we were now on a first name basis, could show up and repair the damage.

"I want you to rip this plumbing apart and replace every line and every fitting. I want water cut-off valves in every room," I said.

He happily smiled and went right to work.

Later that evening I was singing happily as I handed my husband the plumber's bill. Total: One Thousand, eight hundred dollars. Having water: priceless.

I Don't Drink, but?

Now I am of the belief that if something is broken you either bang on it with a hammer, wrap it with duct tape or call a professional. My husband on the other hand is a do-it-yourselfer, which in past years when it comes to the simple things or the mechanical jobs, has worked well. What I've learned is that as we grow older, our patience level grows shorter and what we used to know, somehow has been forgotten over the years. This and homesteading can bring some quarrelsome times and hot tempers as we stumble our way through trying to become self-sufficient.

So, when he told me on this on Saturday morning that he planned on changing the serpentine belt on the truck, I was like, "Oh, okay sounds good."

He has worked on vehicles for many years and I had the utmost confidence in this being a simple, quick and easy job. My day had already been planned. I was staring at two bushel of corn that needed to be canned and put up for winter. I had laundry to do, and also had rocks I wanted to move to start building a stone wall around where my garden would be. We would both be busy enough, I thought, not to get on each other's nerves.

As soon as the sun rose in the morning sky, I was dressed and shucking corn. On the stove were two huge pots of boiling water to blanch the ears then remove the kernels from the cob. I would then raw pack them and set them into the canner where they would stay for an hour and a half, long enough for me to start the laundry. Multitasking had become a way of life for me.

"Honey," my husband yelled from the door yard.

All I could see was his denim-clad legs and butt hanging out from beneath the hood of the truck.

"Yeah?"

"Can you hand me the socket wrench?"

Grumbling, I walked out to where he was and dug through the toolbox in search of the socket wrench. Handing it to him, I turned to go back into the house and resume my own chores.

"Ummm… I could use your help out here," he muttered. Rolling my eyes and biting my tongue, I turned back.

"What'cha need?" I asked sweetly.

"Well, it's a pain trying to climb up and down for tools, could you stick around and hand me what I need."

I thought of the canner slowly chugging away and the corn husks that were making a mess of my kitchen table that I hadn't taken care of yet.

"I guess. Sure," I replied.

For an hour I handed him tools and listened as he spewed out one profanity after another. He banged and grunted, pulled and twisted one bolt and nut after another and finally jumping down from the truck after slipping a wrench and busting open his knuckles. I ran for the band aides as blood ran down his hand. Cleaning the wounds and taping him up, I looked at him impatiently.

"What's the problem?" I asked. "You said changing the belt would be easy?"

"My hands are too big, and I can't get to the nuts to loosen the belt," he complained.

Oh, I knew where this was headed. The last thing I felt like doing on this hot Saturday was playing mechanic. The last time I'd helped him fix a truck had been more than thirty years ago and he damn near dropped an engine on me. It hadn't been pretty, and an angry conversation of divorce had followed. What I knew from many years of experience is that we didn't work well together at all. It was kind of like mixing gasoline with a match.

"Why don't we call Steven? He can help you," I suggested.

My son was a good mechanic and used to his father's impatience when it came to tinkering on the vehicles, whereas I was easy to lose my temper and blow.

"No. We can fix this. I just need your small hands to loosen a few nuts," he replied.

Loosen a few nuts? Oh… I had thought of nuts, but they weren't attached to the truck.

"Okay, give me the hammer, those bastards'll come off," I replied jokingly.

He saw no humor in my reply.

Resigned to the fact that I would be covered in grease, I climbed the bumper and slid under the hood of the truck. Showing me where the nut was that he couldn't get to, he handed me the wrench. Snaking my arm down through hoses and engine parts, laying my cheek on something greasy, I placed the wrench right where he'd told me to and tried to turn it. No luck. I was not strong enough to break the nut loose. I shifted position and looped one knee around the hood release and turned further onto my stomach. Grunting I tried again. Again, no luck.

"Honey I just can't! I am not strong enough," I said and grunted as my husband climbed up behind me and reached over my shoulder trying to grab my arm and give me more leverage… basically, laying on top of my back, my legs in between his spread legs, spooning me in the most intimate of ways.

It was at that moment when we heard the cough behind us and we both looked up to see our neighbor standing there with a grin on his face. Really? Really? I looked at my husband and he looked back at me and I thought of the grunting and groaning we had been doing. I wondered how long my neighbor had been standing there and God forbid, what he was thinking.

"Excuse me, um, sorry to interrupt but do either of you know anything about how to slaughter a pig?"

I am sure my face expressed my confusion as I looked at him like he had twelve heads and horns. What? What? Did he just say something about a pig? And slaughtering it? Really? I am laying under the hood of a truck with my arms deep into the belly of this beast, covered in grease and my hair wound around nasty, greasy hoses because I hadn't thought to tie it up beforehand, getting ready to give birth to a serpentine belt with my husband basically laying on my back and he wants to know if I know how to slaughter a pig?

"Ummm, no man. I don't," I replied as I unwound myself from the truck and my husband.

Climbing down I saw that I was covered in grease, the knee of my pants was torn and in all my contortions I had lost a sneaker. What was with these neighbors showing up at the most inconvenient times?

"Well, I got the pigs ready to slaughter and my helper hasn't shown up yet. Just thought you might know how," my neighbor explained.

Trying to be helpful, I tried to remember what I'd learned about pigs.

"Well, you pretty much just shoot them in the head I think," I replied.

I heard a snort from my husband as he walked toward the house to grab a drink, essentially leaving me on my own to discuss pig killing with my neighbor. I wanted to slap him.

After a forty-five-minute discussion on killing a pig, my husband hiding in the house, my neighbor finally got the hint and left. Back to the truck, the grease and the nuts that wouldn't turn. I couldn't help but wonder about that poor pig and also wondered whether I would hear its pitiful death cries from up over the hill. God, I truly hoped not.

"Okay, I have about two minutes of patience left before I start beating the hell out of this engine or you with a hammer," I snapped three hours later.

We had woven the new belt through a dozen times and it still wasn't right, we had come close to whipping tools and smashing windshields. Murder was in the forefront of my mind and I am sure on his as well.

Neither of us could figure out what we were doing wrong. I was crouched and guiding the belt along the pulleys that he was telling me the belt needed to loop around and yet it still didn't fit around all of them. Angrily I grabbed the diagram and studied it. Over pulley A and under pulley B, around pulley C and behind pulley D…. I looked at the diagram, I looked at the engine and I looked at my husband with fire shooting from my eyes. Why he couldn't have just called our son to help?

"Call Steve," I said. "I swear honey if you don't call Steve, I'm going to throw this hammer right through the windshield," I threatened.

"No, I'm sure he's busy," he replied as he studied the diagram and looked at the truck engine, ignoring me. What is up with men who wouldn't ask for directions or help? Seriously!

"I'm sure he's not. Call him," I argued.

"No, I don't want to. I'll get this figured out," he replied.

Oh, my God my head was going to explode! Feeling my blood pressure soar upward to the nether regions of total insanity, I glared at my husband.

Matches and Kerosene

Now, for those of you who have read my stories, you know, my hubby and I do not work nicely together. When we do it usually ends up with hurled profanity, insults, death threats, and tools (mainly the hammer) flying well aimed at (mainly his) head. I have a hot temper and projects with him tend to bring out the worst in me.

Having said that, some projects just demand that we try to play nice. And building our work shed combo woodshed was just one of those together projects that demanded we do it together. Throughout our first winter on this little patch of heaven, I dug wood for my woodstove out of piles of ice and snow, mud, and rain. So, during those long and dark winter afternoons hubby and I would talk and plan of the dry, enclosed shed we would build so that the following winter we would not be fighting the weather for our winter heat.

May came, and excitedly we began our woodshed project. Me, wanting it done today and he, moving turtle slow. Seriously, this man moves like molasses trying to flow uphill! May comes and with it the black flies. Chewing on me, flying up my nose and down my throat, clogging in my eyes. June horseflies….

Still, just a floor, no walls, no roof. July, long weekend for the holiday, sweating, cursing, fighting heat and hoping I don't murder this man before the weekend is over. One wall up. I'm now beginning to panic because I know fall is just a few short months away and I still have wood to stack, a chicken house to winterize, a garden that will need to be turned and prepared for the following year, a new door to put on the house, plumbing to fix, underpinning to paint and repair, logs to seal and paint on the cabin and canning to start very soon. I have visions of stabbing hubby in the throat.

The end of August I finally lose it and start

lighting a fire under his ass with threats of death, or worse, torture for the rest of his miserable life. We finally get all the walls up. Now at least I can start stacking my wood in the building while we work on finishing up the roof. So, four and a half cord of wood goes in and in record time. By Labor Day weekend all my winter wood is stacked and ready. I feel good, pressure off and now I can start on my other projects.

Sitting on the porch, patting ourselves on the back for the work we've accomplished, we hear a loud crack and boom. I look at hubby; he looks at me.

"What in the hell was that?" I whisper.

"Sounds like something in the shed."

"Well go look!" I hissed.

"You go look! It's dark out there."

Neither of us went and looked. I figured it was a raccoon or bear and really, who wants to run into either in the dark.

The roof. I loaded up the truck with more boards. Our last step and gosh I was so glad it was almost over. As hubby stood up on the ladder behind the shed, I handed up board after board. Something looked off kilter to me, and I just couldn't pinpoint it. Until I looked down and saw, and my stomach sank. The loud bang and crash we had heard the night before were, in fact, the floor ripping away from the walls, collapsing in on itself.

"Ummm honey, Ummm we have a problem."

"What?"

"Problem… problem, we have a fudging problem!"

Climbing down off of the ladder, he looked to where I was pointing. The words that came out of his mouth would have made his daddy proud! (I am being sarcastic here).

Sitting on the front porch, exhausted and defeated, I watched hubby as he walked in circles around the shed looking for a simple solution that wouldn't include tearing the beast apart and rebuilding. My solution as I yelled it across the yard did not impress him.

"Gasoline and a match!"

I got the eye roll.

"Honey, we got to get the wood out of there. I got to fix that floor."

Now, I know the (we) is in fact, me… unless he's got a freakin mouse in his pocket.

"I hate you right now." I hissed. "Give me your knife."

He looked at me worriedly. No, I wasn't going to commit suicide… I was going to commit murder.

"For what?"

"So I can freaking stab you in the 'juggler'." I yelled. Every once in a while, he will come back with a zinger and this time proved to be just one of those rare responses that set my blood to boiling where I truly do think my head will just explode.

Turning mid stride, he smiled sarcastically.

"Honey, violence will not solve this problem."

I whipped my coffee cup at him.

Thus, I spent this week, in the rain, in the humidity, in the mud and chicken poop, unloading four and a half cord of wood. Do I still entertain visions of killing my husband? Oh, hell yeah, but I also want my shed to be finished, so, for now, all I can do is entertain those visions and stow them on the back burner. After all, I still have to load and stack back into the shed, FOUR AND A HALF CORD OF WOOD!

Stinky Situation

Okay, I can now get the dumb ass of the year award. I get up at 4 a.m. and am impatient to get started on my day as the heat is supposed to climb later. Before light hits the sky, I am out in the prison yard, tending to the Bad Boy of the Barnyard and the Little Chicken Mafia with my faithful dog by my side as I work with flashlight in hand. I scatter chicken grain for the little carnivores, shovel chicken poop from the floor and make several trips to the compost pile. I smell like I just took a bath in the duck pool.

Hurricane Jose on my mind as I think with agony of the wood that I still have to stack and wanting it done before the storm comes in. I am done with the poop patrol by 5:30 and it's still dark, murky with fog and Oh, so quiet on the home front. I love the early mornings, it's when I can work and let my mind venture to faraway places.

Hauling out the wheelbarrow I notice my Tara Belle acting strangely and not giving it thought I head to the wood pile and begin loading the wood once again. It seems this woodpile is mocking me. I've loaded and unloaded this freaking thing 4 times now, which is another story for another day.

Calling Tara Belle, thinking she's just spooked by something she smells in the woods; I make my way to the woodshed. The floor is still holding so hubby won't die by my hands today. Tara belle refuses to follow me and damn it, I should have listened to the dang dog.

With an arm load of wood, I make my way into the shed. I see the skunk and that little skunk sees me. Our eyes meet, hers curious, mine horrified. I scream like a schoolgirl.

"Oh, dear God why?" as I back pedal to get the hell out of the way. She squeaks, lifts her tail and Pepe Le Pew sends out a spray of the most noxious, horrible, gagging substance ever created on God's green earth. I drop the wood and run like my ass is on fire and the devil himself is hot on my heels. Tara Belle is already on the porch, and this little fu***** Pepe Le Pu is chasing me! Since when do skunks chase people? Aren't they supposed to run away? Thus today, my shed and wood pile have the most noxious odor known to man and I am at a standstill with loading the wood. Hurricane Jose can have it, and perhaps in all his hurricane fury he will carry that God-awful smell from my yard....

The Barnyard Rumble

The human body fascinates me. It is strong, resilient and this working machine, if working properly, carries us day in and day out. Yet, one wrong move, one simple twist and the body will let you know who's in charge.

As a homesteader my work starts at dawn and all through the day I am twisting, bending, reaching, lifting and putting my body through its paces, often pushing it further than it wants to go. So, you would think something as easy as picking up a lawn chair and moving it would not be a big deal, right? Wrong!

It was a simple task. Pick up the lawn chair, fold it and bring it into the hallway so it doesn't get wet with the storm predicted to come in. And that lawn chair kicked my ass from here to sundown! One simple move, one light task after all the heavy and hard tasks I do, was enough to get me whining like a two-year-old. Seriously.

But the work doesn't stop for an ache or pain. So off to the kitchen I go to whip up some herbal pain relief. And it helped to a degree. Moving like an old, hunch backed lady, I made my way out to the prison yard. Feed, water, clean the pen and let the prisoners out. (As a side note, animals can sense weakness) Holding my breath, moving slowly I went through my morning routine…. Peckerhead was crowing inside the coop wanting his breakfast. The Little Chicken Mafia singing their morning chorus as they pooped out my breakfast. With each step I take, each movement of my body I am reminded that the body is a unique and fickle thing. And today she was just a fickle pain in my back.

Opening the door to the coop, I am greeted by Peckerhead. This rooster stands 2 feet tall but has the nazz of a Tyrannosaurus Rex. His curious eyes are scanning me, wondering whether today would be the day to challenge his hierarchy in this flock. Yes, they include me in the flock, and I am top chick. Sensing my slow movements, I know he was plotting to kill me. I could see it in his eyes and in his body posturing. Oh, no, not today of all days big boy, I ain't got time for this crap. Nudging him with my foot, I move to open the grain barrel. The Little Chicken Mafia are watching me hungrily. I don't know if it's for the grain or if they are craving human protein; they are little carnivores after all.

Peckerhead is strangely quiet and I watch as he circles my legs. I am on high alert. They have sensed my weakened state and I just know he will make his move. Backing out of the coop, I shut the door and go to put grain in their outside dish. Peckerhead meets me in the pen and I know the dance is on.

"Okay, you little flucker… let's rumble…." I growl as he puffs up his white feathers and flaps his wings. Did you know that chickens can growl? Well, now you know, they do! His eyes zero in on mine. His neck feathers fluff out, making him look larger and more menacing. I stand my ground, knowing the day has finally come where he will try to claim the title of King of the Barnyard. I stare at him, challenging him to take his best shot, my booted foot ready to punt him like a football.

I'm thinking the best way to grab that little flucker and flip him upside down. This, I found out, is the best way to assert dominance and keep my place as top Chick in the pen. My back is screaming at me because I just know it's going to hurt a lot worse after this little rumble with my rooster.

"Okay, bring it on…." I growl as I take the offensive and move toward him. I puff myself up, looming large, trying to intimidate the beast. The Little Chicken Mafia are watching me with beady, curious eyes. I just know they want to see me get my ass kicked by him.

I lunge, sending my back into a spasm that nearly brought me to my knees, and he lets out a squawk as he quickly backs up. He's seen my weakness, but he's also smart enough to see that I am still top Chick and still going to kick his feathery little ass seven ways to Sunday, even if it kills me. I keep advancing forward and he keeps retreating. The battle is won before it even gets started. With shame he hangs his head and growls as he retreats back into the coop. The Little Chicken Mafia sigh in frustration. I believe they were behind this little uprising, plotting and planning through the night to overthrow the current Queen (me). I turn and smile at them sarcastically.

"Good try bitches! But not good enough!" I laugh. Today will be a day of ice packs and heating pads. Oh, this fickle body of mine.

High Heels or Shit Kickers

Now normally I share with you the funnier side of living and loving on this homestead, but there are those days… meh? Not so much fun. And today was one of those days. I woke up just plain tired. Physical, mental, and spiritual fatigue. Pushing myself out of bed, I sat on the front porch, drinking my coffee and watching as the night sky gave way to the milky light of dawn. Washed in an unsettledness I couldn't pinpoint; I went about my usual morning routine. Poop patrol, feeding the feathery children of the barnyard and cleaning out the coop. And with each chore I struggled with what was normally very easy for me….

It has dawned on me that as much as I love my solitude here on this little homestead, I also sometimes feel isolated. I get sick of hearing my own voice and need the interaction of others. You see, there are months here on the homestead when I rarely have time to do the girl things… the lunch out, the movies, the girlfriend dates… months of seemingly endless chores. Something always needing to be done and I get so immersed in this life that I forget to take the (me) time I need. Yes, I do run errands away from the homestead but those are quick ins and outs, focusing on what I need to get and quickly returning home.

And today was one of those days as I was running errands that I paused and looked at the busyness and activities of others. The store clerk, the bank teller, the fast food worker. And I wondered… am I missing something? Perhaps I am abnormal living this lifestyle that I do. And these people… they are the normal ones. Should I cash in this lifestyle for a paycheck? Because honestly, my paycheck consists of fresh eggs and vegetables I've grown in my garden, heat from the wood I've stacked, meat from the animals I've harvested…. Should I cash in this lifestyle for human interaction which I sometimes sorely miss? Yes, I do have days such as this. And I know those days will occasionally bring me to my knees with doubt….

It is on those days I've learned… to reach out… to grab my faithful sidekick, my sister, and head out for some normal girl time… to laugh, to eat good food, to be silly and loud and deep into wonderful conversations… because this too is a part of homesteading… knowing when you need the break from the work of daily living, knowing when you need to realize that this very work will wait until tomorrow because your soul/your joy/your connection needs you more right now. It isn't that I don't appreciate this lifestyle… I do with all my being… but sometimes, I just need to step away… to live outside of the Little Chicken Mafia and the Bad Boy of the Barnyard, to forget about loading and hauling firewood or cleaning out the coops, to know it's okay that for one day I can take a step back and do those girly things.

Bump in the Night

Thump….

Hubby asks, "What was that?"

Me, in my head, 'Yup, like I sit up all night standing guard over this bed to see what every little noise and bump in the night is so I can tell you when you ask me stupid questions at 3:00 in the morning. Idiot!'

What I actually say is, "Lucy (our dog) probably jumped down off the bed." Dozing back off I hear another thump and feel a bump under the bed.

Hubby: "What is that?" By this time, I'm ready to start swinging. Those who know me also know that I don't wake up easy… nice… pleasant…

Me: "I don't know, why don't you get up and look!"

Hubby crawls out of bed and goes to make his coffee. Dozing back off I am suddenly woken again… this time by my sixty-pound Tara-belle jumping on my bed. Now I am seriously pissed off! At 3 in the morning I am hitting my best sleep and for whatever reason the universe has decided that I've had enough rest. To me it was just downright cruel.

"What are you doing?" I yell at the dog as she's scrambling across my prone body…. in the kitchen I hear hubby yelling at the two other dogs to knock it off as they whine and bark at the bedroom door. Opening one eye I am getting ready to start throttling animals… and I see what the ruckus is all about… screaming I scramble to jump out of bed as a mouse, being chased by a sixty-pound dog, scampers across my legs.

Jumping up, half asleep I dive out of bed, and in my terror filled haste the blankets and sheets get tangled up around my legs. I yelp in pain as I hit the wall and twist my knee… and as I sit in a heap on the floor, my little dog is excitedly climbing me like a tree trying to lick my face… I live in a zoo….

Tara-belle jumps off the bed as hubby stomps toward the bedroom questioning what in the hell is going on… she blows by him like her ass in on fire and knocks over the fan in the living room then proceeds to wipe out the kitchen chairs and my sewing machine… in essence… hell bent on catching the mouse and destroying my house in the process… and I come flying and limping out of the bedroom screaming and chasing her down before she does any real damage.

The two little dogs are barking and chasing after her excitedly… hoping to get a taste of the mouse that is now running everywhere and anywhere trying to get away from the beasts that are determined to eat it for breakfast… the cat is sitting on the stairs looking at us all and I swear I can hear her laughing… and calling us all a bunch of idiots! Hubby meanwhile is chasing the little girls while I'm desperately trying to tackle the bigger dog….

The mouse got away, I've got a wrecked living

room and the dogs are sleeping on the couch after I threatened them all that they would be going up for adoption if they ever pulled that shit again…. And as I sit outside waiting for my coffee to work its magic, I gaze up at the starry night sky… this is home sweet home… zoo and all.

Appreciation

There is a refreshing chill to the air this morning and as I sit, drinking my coffee, and gazing up at the early dawn sky, I can't help but think of the long list of projects I've still yet to do before the first winter snows kiss the earth. My warm breath comes out in little white puffs telling me that fall is truly upon us. The air is perfumed with the sweet smell of falling leaves and moist soil.

It is surprising to me how far I've come on this little piece of dirt in one year… and it is surprising that although I have been busy with endless projects outside there are just as many projects taking place in my small kitchen. It dawns on me that a lot of homesteading takes place in the kitchen. And to me, this is where the heart of the homesteader is. Cooking wholesome and harvested foods. Providing nourishment for the body as well as the love connection within the family. Knowing that my hands that have prepared the food are not only filling tummies but filling hearts as well as we sit around the table and spend time with each other. The tears that have been shed, the laughter that has echoed around the kitchen table. The holidays and family gatherings, the milk and cookies with my granddaughters, the long and quiet conversations with my sister. The nights sitting quietly and sipping a cup of hot tea as I plan the next day's activities. The time I spend in the kitchen, I realize, is just as important as the time I spend outside.

Throughout the summer I am busy putting up vegetables from my garden, each ripening at various times and it seems to me at some points, all at once although it truly isn't. I am busy baking bread a couple of times a week and drying herbs from my little herb beds. I am busy making medicines to fill the home medicine chest, natures healing gift to battle colds and flu's, to take down fevers and soothe sore throats. To ease achy muscles and to heal wounds.

It is now that I realize that in so much of what I do, I do for the benefit and the growth of my family. So much of what I do is for the nurturing of my family in the most loving way possible. And it is a peaceful and easy feeling that I sit with this morning. Could I have chosen a different path and gotten this same result? I don't think so. Because this is where my heart is… on this little 2 ½ acre plot of land that I am gifted to take care of. It is only being led here to this lifestyle that I have found this peace. And what a beautiful peace it is.

Frisky Critters

In the barnyard, the mighty barnyard, the rooster sleeps tonight…. I bet you're singing along, eh?

But seriously… it never stops on the homestead and some conversations that take place… Lord have mercy.

As a homesteader I am always learning. And a lot of my learning comes from others sharing information. So naturally I've joined many homesteading groups on the internet, to ask questions, to learn new skills and crafts and just because the conversations are always stimulating, often funny and sometimes downright bizarre. With these groups I've talked about types of chicken poop, what to look for in the said poop, types of mites and lice chickens can carry, what a fairy egg looks like, what types of worms to look for with my chickens, what oozes and drips and spurts and crusts over. You know, those gross things that you DON'T talk about over the dinner table. Like chicken vent issues…. I found out is the place where they poop out their eggs, or what signs to look for if a chicken is egg bound and what to do about it….

So, for me the conversation with my husband about our rooster's rather amorous behavior seemed quite normal. Right?

Me: "Peckerhead, knock it off and leave her alone."

You see, I have a very attentive rooster…. He loves his flock. Almost too much sometimes.

Hubby: "Do you suppose the girls find his little dance attractive? Like he's a virtual Don Juan?"

The rooster dance goes like this: A low soft growl as he approaches one of the hens. Said Hen chitters back and moves away. Persistence though is Peckerhead's middle name. He moves back to her and fluffs up his feathers and makes himself appear larger. Bobbing and swirling his head he growls again almost as if he's saying, "Hey look at me. Look how large and handsome I am."

Nope, Hen does not think so and glares at him like she's warning him… "Back OFF buddy, I've got a headache." Peckerhead though isn't ready to give up. He softly and quietly slides up beside Hen again and this time he stands on one foot, hopping up and down, growling and whispering sweet nothings in her ear… all the while flapping his wings.

Me: "Peckerhead! Leave her alone!"

Hubby: "I think she finds it attractive. Do you think she finds it attractive?"

Me: "Why are we having this conversation? I don't freakin know if she finds it attractive! I'll tell you what, why don't you, tonight, when you come to bed you puff yourself up, flap your wings, growl and hop up and down on one foot… we'll see if I find it attractive…."

Hubby: "Mmmmmmm…."

I cannot believe he was actually thinking about trying to impress me with the Chicken Mating Dance….

Me: "Oh, honey, we really need to have a day off of this homestead."

For God's Sake the Mouse!

The almighty mouse! Now being on the homestead there are certain rituals you find yourself doing every day… and they become habits. This log cabin has mice and no matter how I try, I just cannot keep up with the little gray fuzzies. I have three dogs who are great mousers, a cat that isn't. Go figure.

The morning dawned bright and cold. The moon still high as I began my morning chores. I was in a rush, due to leave in the next hour to go visit my brother who is in the hospital, so really not paying attention to what I was doing. Just moving into the ordinariness of my routine. I slipped into the hallway to put on my boots, gearing up for getting the dogs out and morning poop patrol done.

As I slid my foot into the first boot, Lucy Lou, one of my smaller dogs was climbing up my back, anxiously waiting her turn to potty. Hurrying, I slipped my foot into the other boot. Now normally with all the creepy critters around this homestead, I check and shake my boots before putting them on but in my haste... it was the one step I failed to do. So, when I encountered something wiggling against my foot needless to say, I freaked.

I screamed like a schoolgirl and struggled to pull my boot off while doing the hokey pokey in the hallway.

Lucy Lou... startled by my scream proceeded to piddle on the hallway floor, which I stepped in with a stockinged foot. In all my dancing and frantic kung-fu moves I noticed not only did I have a gray, beady eyed fuzzy mouse in my boot but now it was clinging to my sock for dear life!

I had visions of it crawling up my pant leg… scrabbling against my bare leg to reach places in the nether region… screaming, squealing and kicking wildly, I did everything in my power to dislodge that little bastard… praying, Dear God why? Why?

The dogs, now all excited were clambering

around my legs, the cat eyeing me fiercely waiting for her turn to pounce. It was a freakin horror show going on in my hallway! Somehow in all my frantic, frenzied moves I managed to dislodge the fuzzy little critter who I think was as terrified as I was… at least that was what the expression on his little face indicated.

As he flew from the end of my foot across the hallway my large dog, Terra Belle, chose that exact moment to bolt between my legs and give chase, efficiently and expertly knocking my feet out from under me… and I landed on the wood box under the hallway stairs. Stunned, I sat there, eyes searching frantically for the mouse, dogs crawling and jumping on me and a sock full of dog pee.

Oh, how I love this homesteading life.

Reflection

I wake to the sound of the rooster whom we have named Peckerhead, crowing his morning call. The landscape is still covered in darkness and a full super moon is playing with the trees of the forest, creating shadows and light to entice me. I find my own personal shadows, for one brief moment, lifted as I walk among the frost covered grass and my eyes drink in the beauty of this early morning. I am graced with how grateful I am to be here. No city lights to wash out the brightness of the moon. No cars or busy noise to drown out the silence of this perfect dawn. An occasional call from Peckerhead reminds me that daylight is just moments away.

Yes, there are so many hard things about this lifestyle I choose… but there are also these moments… moments that capture me and take my breath away. The bumps, cuts, scrapes, and bruises seem not so important… a distant memory. The bleeding hands from working hard and the tumbles I take from tripping over my dogs, my chickens, and my own clumsy feet…. Then there is just this… the beauty of each day. And as I watch a loved one struggle… I wish upon him just this peace… just this moment… just this gift. I'd love to bottle it and bring it to him, bathe him in just this… comfort, shadows, light, gentleness.

Mud, Mud and More Mud

There comes a time, every so often, that I find myself needing to get away from the everyday demanding work of the homestead. Where the clucking of the Little Chicken Mafia and the quacking of the Little Nasties are just a breath away from making me crazy. It is those days that I head off into the woods to hike and appreciate the surrounding peacefulness… and as always, I have my faithful Tara-belle by my side. She protects me, alerts me to any big bad things that may be lurking around unseen by my meagre, pitiful human eyes. And I love my dog… my faithful companion… my sidekick in every adventure.

Walking along, letting my shoulders relax and my gaze not settling on anything in particular… I am not paying attention to where my fur girl is….

That is until I hear the splashing! That is until I remember the swampy mud hole that should be frozen over… but no… it is not. I ask you, have you ever tried to wrestle a sixty plus pound dog hell bent on bathing in a mud hole?

Dear God, it isn't pretty! I call to her… and blatantly she ignores me. I curse at her, threaten her, plead with her to get the fu*** out of the stinky swampy water.

She turns her head and gazes at me with those deep chocolate eyes of hers… daring me .... challenging me… taunting me….

And she knows… she just knows that I do not want to sink knee deep in mud and muck with my sneakers on to drag her sorry ass out of the swamp. I swear to God in heaven I saw her smile smugly at me as I edged nearer and nearer to the putrid stench of the swampy water, muttering the whole time about how much I could so give her up for adoption at that point.

'Don't you make me come in there after you girl!' I heard myself hiss.

Too late did I see her bolt… straight at me… knocking my feet from under me… sinking me ass deep in mud, in the stink, in swamp water…. and in my haste, I turned to try to grab her collar as she sailed past me… adding insult to injury, dragging me further into the mud, into the wet, slimy leaves… all the while wagging her ass like it is the best game in town….

Tara-belle one… this muddy girl… laughing her ass off… zero!

Oh, this afternoon hike turned out to be the best part of my day!

Sigh- Laundry Day

Today is laundry day! Quite the process in this house as the washer I have handles only five or so shirts or if I'm doing jeans, only two pair at a time. Yes! I do laundry quite differently than you. I have to run the drain hose into the toilet; I have to bucket water into the washtub to wash, drain, then bucket water again to rinse. Then I have to wring each piece of clothing out and transfer into the spin side. Normally this is a time-consuming but very easy process.... Except for this morning.

This morning my mind was having a rat race-monkeying around and chasing shadows. Those little buggers can distract me to no end.... I swear, where others complain of monkey mind, mine is a pack of freaking drunk chorus girls and when they get to raising hell, all bets are off for any type of intelligent thought.

Back to the laundry. Stumbling around half asleep, four a.m. comes early, trying to drink a cup of coffee and rush through washing the clothes, I utterly and completely forgot to run the drain hose into the toilet....

My bad! Doing dishes, I'm listening to the washing machine run through its cycle... quickly I switched the dial to drain and rushed back out to the kitchen... only to hear the sound of water gushing across my bathroom floor... sudsy, gray, dirty wash water.

Now, you know with the animals (three dogs) that I have, this house is often as rowdy as turning a kitten loose in a dog park.... Can you just imagine it...? Water flooding out the bathroom door into the kitchen, me screeching for the dogs to get the hell out of my way as I rush to and fro in my blue Eeyore bathrobe, getting soaked clean through, tripping over three dogs to grab towels to mop up the mess and dogs clambering excitedly up my back and licking my face (the little one happily humping my leg) as I'm on my hands and knees mopping up the floor before the water hits my living room carpet….

Dear God in heaven you don't know how much I'd give right now for a normal washing machine and dryer… but in this cabin there are no hook-ups nor room for either. And truth be told, I don't mind my little washing machine… and drying the clothes in front of the woodstove on a wooden drying rack does not bother me so much….

So, this is how I started my day today. And I wonder? How is your day going? From the shits and giggles here… have a wonderful day.

Oh! To Town We Go!

There are times when I am so hurried, so focused on the chores that need to be done, that it leads me to being forgetful… and often this forgetfulness can lead to some horribly embarrassing moments.

On this day I had to make a trip to the grocery store. Bread making was on the agenda and I found myself out of yeast. So, with this on my mind, I rushed through my morning chores with the intent of getting to town and back before the day got away from me. And hubby decided to go along for the ride.

H: "What is that smell?"

Me: "Mmmmm… I don't know. I don't smell anything."

Silence.

H: "What is that God awful smell?"

I looked at him and shook my head.

M: "I have no idea! I don't smell anything!"

Silence.

Me: "Did you step in dog poo? Maybe that is what you are smelling."

H: "I don't think so. He lifts each foot and examines the bottoms of his boots."

Me: "I don't know then."

H: "It smells like something freaking died in here." (As he is now gagging and cranking the car window open to hang his head out.)

Me: "You are such a friggin drama queen! I don't smell anything!"

At this point I am so irritated with this man that I am seriously thinking of pulling over and bootin' his ass out of the car.

Me: "Keep it up and you'll be walking home."

H: "I'd rather walk than smell this stench all the way home."

Me: Smiling. "I can make that happen dickwad."

Silence.

H: "Oh, dear God I am going to puke!"

Me: "I don't know what you are smelling… shut up already."

Getting through the grocery store, I grab what I needed and made my way back home. It was a stony, silent ride. Him hanging his head out the window and me staring straight ahead… until I started laughing…

so… hard… that… tears .... were… running… down… my… face!

H: "What in the hell is the matter with you!"

Me: Between guffaws and big huffs of breath… "I forgot to change my shoes!"

H: "And?????"

Me: "My barn shoes…." Laughter and snorts as I choke it out, that God awful smell is the heater blasting on my barn shoes…. "Which .... by the way, you are smelling…. duck poop… chicken poop and only God knows what else…."

H: "I am going to friggin kill you!" As he proceeds to keep his head out the window….

Are you kidding me?

Chicken butt. Now, I can seriously and honestly tell you I know more about chickens than I really care to. But it hadn't always been this way. I was raised on a farm and yes; we had chickens but the most contact I ever had with them was collecting their eggs and scattering their feed. So, when I got chickens on my own little patch of dirt years later, I was starting out with zero knowledge of chicken care taking. I knew to water and feed them, but that was about the gist of it.

And as I turned to Facebook chicken groups to get information, I couldn't help but wonder what kind of crazy chicken people I had taken on as friends. I mean seriously, these peeps put me through some pretty gnarly paces….

One August morning I walked out to the coop to do my usual morning routine… collect the eggs, feed the Little Chicken Mafia, turn them loose with fresh water. No big deal, right? Wrong!

One of my bitchiest little prisoners didn't seem quite right to me. She was sitting on her nesting box, head drooping and looking pretty sickly. Panicked, I picked her up. Had something hurt her during the night? Had the Bad Boy of the Barnyard beat her up? Checking her over, I could see nothing that would warrant her behavior. So, setting her back into the nesting box I ran into the house and jumped onto the internet to ask my Chicken Peeps for advice. The conversation flowed something like this:

Me: "Help! One of my hens is not looking good this morning. She's droopy and sickly looking."

Reply 1: "Is she wounded?"

Me: "Nope, I checked her from beak to toes… no wounds."

Reply 2: "Mites… check her for mites. Check under her wings. Also check her crop."

Me: "What's a crop?"

Reply 2: "Under her neck, just near her breastbone. If it's swollen, she may have a swallowing problem as it could be impacted."

Me: "Okay. I'll be right back. Don't leave me." (Yes, I am whining.... I am a needy person.)

Running out to the coop I flip birdie off of her nesting box and proceed to look under her wings, on her tummy, around her crop. Nothing stands out to me other than I am pissing her off.

Me: "Nope, don't see any bugs and crop looks normal."

Reply 1: "Is she drinking and eating?"

Me: "Nope."

Reply 2: "Check her vent."

Me: "What in the hell is her vent?"

Reply 2: "Her butt hole! She may have had a fly strike... and it could've brought on maggots."

Me: "Ewwwwww… no! I am not going to be poking around my chicken's ass!"

Reply 2: "You have to. If that is the case, it could kill her."

Me: "Are you freaking serious?????"

Reply 2: "Tee-hee… yes."

Drawing in a deep breath, I run back out to the coop and lift birdie girl again. This time I tip her upside down so I can look at this new part of anatomy called a vent. I am really pissing this chicken off as she proceeds to squawk and peck at me… and her grumblings are drawing attention from the Bad Boy of the Barnyard who has decided to barge back into the coop to protect his lady….

With a toe nudge, I scramble his ass back out the door and close it so I can (what feels like to me, chicken molesting) get a closer look at the chicken's vent. Seeing nothing out of the ordinary… no maggots, no broken open areas, nothing.... I run back into the house.

Me: "Nope… no maggots or broken open areas… do not ever ask me to do that again! That is just wrong!"

Reply 2: Laughter ensues…. "Sorry, but it had to be done."

Me: "So now what… she's looking worse by the minute?"

Reply 1: "Did she lay an egg this morning?"

Me: (Trying to remember how many eggs I

collected) "Ummmm, I don't think she has yet."

Reply 1: "She may be egg bound. If she is, you have to help her get that egg out."

Now I am thinking about what I have gotten myself into with these damn chickens! I'm almost afraid to ask what I will have to do to get the chicken's egg out of her butt.

Me: "And how can I tell if she is egg bound?"

Reply 2: "You have to feel around her vent area and see if you feel something bulging and hard…."

OMG! What is with you people and the vent area? Looking for something bulging and hard? Really? My mind just does not want to go there! Rather than the Dog Whisperer, I feel like I am becoming a freaking Chicken Molester!

Me: "Okay, I'll do it. Don't leave me, I'll be right back."

Running back out to the coop I flip chicken girl over and amidst the frantic kicking of her legs and the indignant squawks I start feeling around her vent and lower abdomen. And yes, sure enough I feel something bulging and hard. She is egg bound.

Me: "Okay, I think she is egg bound… now what?"

Reply 2: "Oh, no, you have to get her to pass that egg or she will die."

Me: "How do I do this?"

Reply 1: "You have to set her in a bucket/bath of warm water and massage her tummy near the vent area. Also try to force her to drink a bit of water with a Tums crushed up in it for the calcium…. you will have to keep giving her butt baths every few hours until she passes the egg, or you will lose her."

Butt baths? Calcium? Oh, good God! How in

the hell does an egg get stuck in a chicken? Really?

Me: "Okay, I'll give it a whirl. If the egg doesn't come out… what then?"

Reply 2: "Well, then you can put on a rubber glove, coat it with Vaseline and very gently go fishing for the egg and try to coax it out…."

Me: "Oh, hell no! Nope, nada, ain't going there!"

Well thank you God, after a few warm butt baths… rubbing her tummy… she passed an egg the size of an ostrich egg… a double yolker to boot… because to tell you honestly my friends, I was not going into the nether regions and pulling that egg out… it just wasn't going to happen… my resume does not include (chicken proctologist).

No Sleep on the Homestead

After working hard all day, the thing you look forward to the most is a good night's sleep. Is that too much to ask? I mean seriously….

Evidently it is.

The night went something like this.

I crawl into bed, snuggle down in the warm blankets and sigh tiredly. Everyone is sleeping already, even the mouse.

I close my eyes and begin to drift off to that

space where my body and mind relax, and I am just on the edge of melting into the night.

Bang, Bang, Bang… startles me and brings me up out of this Oh, so peaceful place. Kitty girl is banging the cupboard doors.

Me: "Kitty girl! Knock it off!"

Sweet silence….

Bang, bang, bang! Growling, I turn over and screech!

Me: "Kitty girl…. Knock it off! Don't make me come out there!"

Silence… and I can just see in my mind's eye that she's plotting to bang those cupboard doors again. Hubby lying beside me grumbles.

Bang, bang, bang! And Lucy Lou, my little fat fur girl jumps down off of the bed to go see what's going on.

Me: "I am going to freak strangle you!" Jumping out of bed, I stumble tiredly into the kitchen to give chase to the cat… and she flies up the stairs and peers down at me from the top step… a mischievous glint in her green eyes as I yell and threaten to make a pair of fur mittens out of her miserable hide. Tiredly, I crawl back into bed.

Bang, bang, bang, crash! Terra-belle decided to join in on the midnight mayhem… jumps off of the couch, slides into the kitchen table, knocking over chairs trying to get to the cat, and the mouse the cat is terrorizing…. Lucy Lou decides she too wants to join the fray and jumps down off of the bed yet again.

Me: "Oh, My God! Just shoot me!"

Hubby: "WTF is going on!"

Me: "Cat, dogs, mouse…. I don't know."

Hubby: "I have to get up in three hours!

Really?????"

Me: "Shoot the cat!"

Hubby: "I ain't shooting the cat."

Me: "Then shoot the dogs!"

Hubby: "I ain't gonna shoot our animals!"

Me: "Then freakin shoot me so I can at least get a moment's peace and quiet!"

Silence…. I really think at this point he might have been considering it.

Drifting off again, I think… ahhhh… sleep.

Bang, Bang, Bang!

Hubby jumps out of bed… me thinking he's going to break bad on the fur kids… but instead starts moaning and groaning… as a leg cramp turns him into a drama queen… and he acts like he's dying….

Me: "Oh, for the love of God! Want me to amputate?"

Him: "Shut up and go to sleep."

Me: "What? With you jumping around like a one legged drunk in an ass kicking contest, the cat banging the cupboard doors, Lucy Lou jumping up and down off of the bed like it's her own personal trampoline, and Terra-belle rearranging the kitchen furniture you actually think I can sleep? I'm moving out! I swear tomorrow morning I am packing my bags and moving!"

Him: "Oh, no you are not… you ain't leaving me with this freakin zoo…."

Me: (laughing)… "Then just please shoot me, shoot them, shoot something!"

Illusions

Okay, so I have a romantic vision of what a snowy day should be: Me, sitting at my desk, writing. Studying my next herbal lesson and making another lotion, potion, tincture or tea, baking, cooking, reading a good book curled up on the couch by the fire.... But in reality, my day sorta goes like this:

Freeze fingers off as I carry a bucket of water to the Little Nasties and the Chicken Mafia, slip a few times on the snow and fall on my ass... throwing said bucket of water high up in the air... make my way through the chicken poop to feed the little killers... fall on my ass and break five eggs in my jacket pocket... get back into the house to find out that the wood box is about empty... go to the woodshed, slipping and sliding all the way while Tara-belle is doing her damnedest to wipe my feet out from under me... load the wheelbarrow only to find it's too heavy to push through the snow.... So now carry the wood a few sticks at a time to load the wood box.... What happened to my romantic visions of the artist, writer, herbalist and homemaker?????? Dashed... dashed I tell ya....

No Pretty Eggs

I have to tell you; I pity anyone who marries into this family of mine. My advice to them is to just hang on, it's gonna be quite the ride. Now, having said this, I will tell you, my sister and I share the same attitude, the same witty, off colored humor and the same outlook on life in general… and every morning we talk on the phone and sometimes these talks just leave me chuckling for the rest of the day. The conversation went like this:

Me: "I have another dozen eggs for you… a mix of chicken and duck."

Her: "Good, I'm just about out."

Me: "Kay, I'll drop them by later."

Her: "Cool but I think Rick won't be eating the eggs anymore."

Me: "Why?"

Her: "Because one of them had some shmuck on it… grossed him out."

Now, anyone who raises chickens know their eggs don't always come out pristine… sometimes they are messy… covered in yuckiness. But they also know that there is a natural coating on them called Bloom, which protects the eggs from bacteria and such. So, most know not to wash the eggs until just before using them in order to not wash away the Bloom. If they are really messy, I generally will wipe them down with a paper towel and call it good but otherwise they come directly from the chickens' butt to my carton in the fridge.

Me: "What did he say?"

Her: "He said, Ewww…. Where did that egg come from?"

Her: "The chickens butt."

Him: "What is that on it?"

Her: "Smuck…."

Him: "I'm not eating that egg."

Her: "Do you think the eggs from the store come out clean and white? No, they get washed, just like I will do with this one you moron!"

Him: "That is just gross… you ain't feeding me that egg."

Her: "Well I would have washed it first idiot!"

Him: "I want store bought eggs."

Her: "Then go to the store and buy stupid store eggs… These are better for you but hey, what do I know? Do you honestly think some little magic egg fairy pops store bought ones from her pristine ass? God bless!"

Me: "So, did he eat the egg?"

Her: "Yup!"

Me: Laughing… "Way to go sister!"

Personal Space?

I woke up this morning with grand plans… and rolling out of bed I hit the floor running. Grabbing my coffee, I held it in one hand while busily putting away dishes with the other as I wanted a clean countertop to make Whoopie Pies which I planned to bring to my brother later in the morning. The dogs were circling my legs as they always do when they sense it's going to be a baking morning… hoping for any little tidbit to fall from the table so they could scoop it up hungrily like a bunch of little beggars. The conversation sounded like a parody of chaos….

"Lucy! Get out from under my feet!"

"Tara-belle don't you dare jump up on that counter!"

"Sadie! For the love of God! You trip me one more time girl and I won't be responsible for what I do!"

Amidst all of this chaos as I am shoving the first cookie tray into the oven, I hear a crash… and turn to find five cases of canned broth, canned vegetable, canned soup and beans scattered across the kitchen floor… smashed… glass shattered and foods and soups leaking everywhere… and my three dogs diving to scoop up anything and everything from green beans to beef stew thinking this is their lucky day….

Now, I am a very patient person… but this fiasco tipped the scales from patience to turning me into a complete and utter madwoman….

Dropping my cookie tray, I screeched.

"No! Tara-belle you move your ass… Sadie… Lucy… get the HELL out of my kitchen!!!!"

Dogs scattered and my soggy slippers are slipping and sliding through the goo to chase them away from the glass… from the puddled stew which is now running across the living-room carpet… and my first batch of cookies is burning in the oven.

Taking a deep breath, my blue Eeyore bathrobe dragging through the soup as I am on my hands and knees mopping up the mess, I am mourning all the work done that is now scattered across the floor… days of prepping vegetables, hours of simmering stews and soups, days of standing over a hot stove canning for winter storage… all of it in a puddly mess that I am scooping up with a towel while watching my darling little girls trying to sneak back into the kitchen to circle me like a bunch of ravenous wolves…. Turning I hiss:

"I just dare you!" As I glare at them… "I will make each of you into stew and can your asses!!!!" I yell as I hold up an empty canning jar…. "See, I have the extra jars now to do it!!!!"

It is sad to say that this all happened before 6 a.m….

Mmmmmmm… I can't help but wonder what the rest of my day is going to be like.

Who Knew?

Woke up early to start the laundry, which in itself, is an all morning process. Clothes scattered all over the bathroom floor, wash tub filled with sudsy warm water, me looking at my ratty Eeyore bath robe and wondering whether I should peel it off and toss it in… but damn it's cold and I just don't want to take it off….

The phone rang, and it was time to go…. Unplugging the washer, I quickly dressed and ran out the door, leaving everything where it was….

Later in the afternoon I called my son.

"Hey, can you stop by and let the fur-girls out to pee, water the Little Chicken Mafia and the Little Nasties and close them up for the night? And add wood to the stove?"

It was going to be a very long day.

Little did I know that my son and daughter brought my Charlie (my five-year-old granddaughter) over to help. And how much this little girl pays attention to all that Grammy does. My daughter, the next day, was telling me of how this five-year-old took charge at Grammy's house.

"No Aunty Ashley, that's not the way Grammy does it! Let me do it." She said as she flapped her arms and herded the chickens into their coop. Picking up the board, she squared it up to the little doggy door and sealed it shut. Turning, she then went to the Little Nasties pen and with arms wide, herded them into their coop… she was large and in charge!!!!

"Now we go inside and close their door." She instructed. How many times she and I have done this together…?

"When you go in Auntie Ashley, watch out for the Peckerhead." She warned, parroting the words of warning I had given to her so many times.

Inside the house my son was stoking the wood fire.

"Daddy, you are not doing it right!" Charlie says…. "You pull out the knob at the top to open the air, you push it in to close the air...."

When it was time for them to wrap up….

"Daddy, we gotta lock the girls in…." and taking charge she instructed Lucy and Sadie back into their crate and promptly locked the door.

As my daughter was telling me this… how this little Bossy Bess took charge, I had to laugh…. For those little things I am teaching her, those little details she paid attention to when I hadn't thought she was, has turned my little Charlie into a take charge little homesteader…. She doesn't mind the duck or the chicken poop, she doesn't mind the Peckerhead rooster who is almost as big as she is, she doesn't mind getting her little hands dirty digging in the hay and nests to gather the eggs….

And proudly she tells me:

"Grammy, I got my muck boots on, so I can help you."

Stew Anyone?

With all the sadness this past few days, it is this homestead that reminds me life does go on. And with that, my Peckerhead reminded me loud and proud this morning. He is such a douche sometimes… but I gotta love that little beast for who he is….

I decided this morning with the weather so brutally cold that my Little Chicken Mafia, Peckerhead (Bad Boy of the Barnyard) and the Little Nasties needed some tender loving care… So, up at 4 a.m. fixing them mashed corn… a nice warm porridge to feed the little buggers. Stumbling outside in the dark, I retrieved their dishes from the frozen coop…. and clumsy me, cut my hand open on the ice trying to break it out of one of the water bowls… swearing up a storm, me and my bloody hand, dripping drops across the pristine white snow, as I made my way back into the house to slap a band-aide on the dang thing before I bled out… hubby looking at me and just shaking his head.

"What!" I growled.

"Not a damn thing." He replied… smiling. He knows me well for my clumsiness.

Scooping warm corn mush into the bowls I made my way back out to the coop where I was met with a chorus of clucks, quacks, and crows…. Not paying much attention, scooting the hungry carnivores out of my way with the toe of my boot I set the corn mush down… and all the little girls attacked it…. Still bent over I opened the door to the Little Nasties coop and scooped up the eggs… and as I stood up… Peckerhead chose that exact moment to fly down off of the roost.

Have you ever had a twenty-pound rooster try to take your head off? I swear to God he did this on purpose. And I, with arms flailing, feathers in my hair, and hay in the hood of my jacket, taking a dive for the chicken shit covered floor….

I WILL HAVE HIM FOR STEW!

Take the girl off the farm

On this farm, with this lifestyle, I don't often get the chance to dress up or wear make-up…. I mean really, why bother? It's not like my Little Chicken Mafia or the Bad Boy of the Barnyard or even the Little Nasties care…. As long as I feed and water them, they could probably give two poops whether I dress up or go out stark naked….

So, on this morning I decided that I was going to paint my nails, slap some make-up on face…. I had places to go and people to see and damn it I just wanted to look decent for a change… not like I had just stepped from the farm….

So carefully applying my make-up… I mean I did it up right! Skin firming lotion, moisturizer, foundation, and cheek color, bold and sexy eyes…. I smiled in the mirror as I thought, "Dang girl, you clean up nice!" Feeling pretty, I went off to the prison yard and performed my morning chores… poop patrol with the dogs, feeding the little buzzards, carrying pails of water… collecting eggs… be extra careful not to fall, slip, stumble or get attacked by Peckerhead…. I congratulated myself on a successful morning without my normal mishaps.

Getting in the car, hubby and I set out to do some shopping…. I was way behind, and Christmas was creeping up on me fast.

Hubby: "What's that shit all over your face?"

Now this question really hurt my feelings… the one morning I take the time and effort to look good and he had to be insulting… so… in a fit of temper, madder than a Chihuahua on steroids, I tore into him… how dare he? The big DF (and I'll let you figure out what that stands for) had the nerve to sit there and criticize my make-up!!!!

Me: "Oh, thank you asshat for that wonderful compliment!" I snarled.

Him: "No, seriously, smuck on your face."

Me: "Screw you! I think I look pretty damn good! God! I really hate you sometimes, do you know that?"

Him: "Honey, you have poop on your face… I am not kidding."

Me: Glancing in the rearview mirror. "Oh, God!" I did have something brown and dirty on my face, and actually, a lot of something brown and dirty streaked down from my cheek to my jaw. The only thing I could think of was that I must've brushed my nasty glove against my face and left this streak of goo behind.

Pulling off to the side of the road, I pulled some

wipes from my purse and grabbed my bottled water. Scrubbing at my face, I proceeded to wipe the chicken and duck poop away… and most of the make-up on that side, leaving me looking like the Joker from a bad Batman movie….

I started giggling and hubby cut me a sideways glance and rolled his eyes.

Me: "Well honey, I guess you can take the girl from the farm, but not the farm from the girl…."

With half a face of make-up, I drove on. My Christmas shopping was going to get done, come hell or high water, come chicken poop and me looking like a two-year-old had applied my make-up… it would get done. Today I set a new fashion trend…. I wonder if anyone else will notice?

Frostbite and Feathers

The bitter cold has finally lifted up here on the hill and I can take a deep breath again. It feels like a whole-body sigh of relief. Everything is harder in the bitter cold. Everyone is hunkered down and movement underneath the layers of clothing is awkward and stiff… as if we can ward off the chill that penetrates our bones.

Setting out, I entered the coop to feed and water the Little Nasties, the Little Chicken Mafia and lastly, the Bad Boy of the Barnyard. Trudging through snow that seems to me to be a growing evil entity of its own, I juggle a pail of water, a lantern and a stash of goodies for the animals. Even before I reach the door, I can hear them stirring around… the girls squawking as they poop out their morning gift of eggs, Peckerhead crowing happily and the Little Nasties chattering quacks as they demand the Queen let them loose from their deep hay bedding. These are all sounds that set my heart to ease… they've made it through the last cold night.

Entering the coop amidst the flurry of hungry and thirsty prisoners, I handle my chores in routine fashion as the carnivores crowd my feet in their hurry and haste to get to the food and water… all except Peckerhead. He stands regally on the roosting pole and peers down at me… challenging me to make one wrong move….

It is a battle of the wills between this mini dinosaur and me. And every day we dance this dance. I peer at him and he peers at me. Silent communication.

Me: "Don't do it, you little flucker!"

Him: "Someday lady, I'm gonna kick your ass! You know this right?"

Me: "Bring it on… I am more badass than you are."

Him: "Soon, soon lady."

In our ongoing posturing, I notice something different about my handsome but dangerous little beast. Stepping closer I hold the lamp up to get a better look and he emits a warning growl… yes, roosters can growl.

"Damn!" I mutter as I gaze at his face. His waddle, which is usually bright red, has white/brown patches on it. Peckerhead was suffering from frostbite on his waddle.

"Okay big boy. Game on as you know what I am going to have to do right?" I say quietly. He peers at me with disdain. Almost as if he's telling me to Fudge Off.

"I don't care! It has to be done so suck it up!" I hiss.

Now, what I am going to have to do is get some Bag Balm and rub it into this mean rooster's waddle. That means I will have to pin him down, avoid his beak and claws and rub the goo as best I can on him. All the while I know the Little Chicken Mafia will be behind my back plotting and planning my death as they circle in closer and closer. It is dangerous times in the coop my friend… and there can be only one winner.

After retrieving my can of Bag Balm, I trudge my way back into the coop. I am on alert and my adrenaline is running high. Peckerhead is still perched on the roost. He knows I've come just for him and he growls a warning.

Being as quick as I can, which isn't fast at all due to all the layers of clothing I am wearing, I lunge at him and he jumps from the roost to the nesting boxes to get away from me. I chase him, hot on his fluffy, feathery little ass trying to grab a handful of feathers.

He moves quickly, turning on me and hisses. I lunge again and trip over the water dish and reach a hand out to break my fall. If I go down, it's all over… he'll jump on my back and probably give a good go at pecking me to death. Round and round we go, me chasing him and him squawking and pecking and lunging at me.

The Little Chicken Mafia are squawking and screeching as they witness the epic battle… the battle of all battles between The Queen of The Prison Yard and the Bad Boy himself, taking bets on who will win. He, hissing and growling and losing feathers left and right, me yelling and threatening as I slog my way through chicken poop and hay, covered from hand to shoulder in both. I can hear the Little Nasties out in the pen shouting encouragement…. PS. They do not like Peckerhead so I think if it came right down to it, they would have my back. At least I hope they would.

An hour later I exit the coop triumphant. Covered in chicken poop, hay and feathers, breathing heavily, I look up at the sunny skies and thank the God of Homesteading that I am not missing any fingers or an eye, that I am not pecked bloody and lastly, that I didn't end up killing that cantankerous rooster…. And I pat myself on the back; yet another job well done.

Dancing in the Dark

This morning is a gift, the type of gift that makes me grateful that I am alive and as I sit in the darkness on my front porch, I watch the fog roll across the expanse; I sink blissfully into the unusually warm temps. We have a kiss of spring in the middle of January and this homesteader is not complaining….

With the cold and stormy weather of late, there has not been much activity outside or in the coop so this morning, before the light has even risen I walk through the murky fog, mist coating my face and my boots sinking into the soft ground where the snow has melted and there is dirt under my feet. Lamp in hand I listen as the Little Chicken Mafia, the Little Nasties and Peckerhead stir about. Grabbing my pitchfork, I am happy to be able to tackle the mountain of poop and wet hay which has thawed over the past few days….

Am I insane to be this happy about cleaning out the coop? To be stinking of duck and chicken poop? Oh, hell yeah and only another homesteader would understand this joy. Opening the door, I am greeted by squawks, clucks and quacks… even Peckerhead is in a good mood this morning as he crows loudly from the roost above. My feet sink into the soggy mess of hay and poop as I deliver the prisoners their breakfast and collect my own…. 5 fresh eggs await my eager hands as I gently move the girls out of my way.

It is still dark but I leave the coop door open for fresh air as I dig in and start scooping up soggy hay, soggy poop and God knows what else that had previously been frozen in a hill the size of a mountain inside of the little prisoners house…. It is a wonder they haven't gathered together, hired a lawyer and thrown a class action lawsuit against me for these miserable living conditions… but I explain to them every day that I am doing the best I can.

After loading the wheelbarrow full several times, I dump it in my compost pile…. Black gold for my garden come spring. Forgetting to close the door, I hear Peckerhead coming up behind me… and turn quickly. He is loose, and it is dark.

Man! Giving chase, I run him through the door yard, around the trees and over snow piles… a rooster loose in the dark is asking for predator trouble.

"Get over here you little Flucker!" I yell as I hot foot it behind him and him staying always one step ahead of me. I am swearing, I am sweating but mostly I am worried that my handsome little beast will take off into the dark woods and become a meal for some hungry animal… damn these early morning calisthenics….

Finally catching him, I deposit him back into the coop… we are both breathing hard and he is now glaring at me…. I think I ruined his good mood. Shrugging it off I inhale deeply the sweet smell of fresh hay as I lay a thick bedding for all the prisoners…. This work makes me feel good, makes me feel whole. Yes, the cold makes things so much harder on the homestead, but when the warmth with its fleeting gift surprises me, I can't help but remember and give thanks for what I have.

Barnyard Rumble

You have all read about my handsome, cantankerous, Bad Boy of the Barnyard, Mr. Peckerhead. Now this rooster was once meant for the stew pot, freezer camp, roaster but to be honest this Warden does not have the heart to follow through with her threats so thus he lives a happy life as the protector for his girls and the nemesis to his keeper. And this is okay for it keeps his and my life interesting as we do our daily dance of existence… him trying to stay alive and one step ahead of me, and me trying to keep myself from getting maimed by this handsome dude. So, on this day, when the bad ass of the barn yard got his ass handed to him by the Little Nasties, I can honestly admit I was more than a little tickled.

Normally I keep the prisoners separated… the ducks in their own pen and the Rooster and Chickens in theirs. But today the mountain of poop needed to be chiseled from the icy pile and removed. And in doing so the prison warden, me, haphazardly left access for Mr. Peckerhead to get into the same pen as the Little Nasties.

And all hell broke loose! And the gang fight began! Fluffy, feathery criminals circling each other, throwing taunts and posturing. Shivs came out from under fluffy feathered wings and the dance of survival ensued.

The Little Nasties, beasty in their own right, circled the Bad Boy who was warily growling and puffing himself up to appear larger and more menacing. Amidst the squawks and growls and angry quacks I hurried to get into the outside pen to break up what would more than likely turn into bloodletting.

As I struggled with the outer metal door, I saw Mr. Peckerhead backed into a corner by the two Nasties. I yelled as I fought with the door. These Nasty little babes would rip my handsome boy apart given half a chance. His eyes met mine through the chicken wire fence. I could see he would not go down without a good fight. I could also see that he would go down. He knew it, I knew it.

I watched in horror as one of the Nasties had her beak latched onto Mr. Peckerhead's throat… and he was using his claws and spurs trying to get her off…

the other Nasty was biting at his butt, pulling feathers out by the beak full. And the damn frozen door would still not open for me to get in and break up the fight.

It was going to be a massacre if I didn't get them off of him soon. Squawking, growls and quacks greeted me as I finally waded in knee deep, throwing punches, kicking and threatening the stew pot for all involved in this prison riot… one of the Nasties bit me as I pried her beak from Mr. Peckerhead's throat… the other Nasty attacked me from behind… it was survival of the fittest amidst the duck poop, frozen swimming pool and food dishes… and I was determined that Peckerhead and I would survive.

Grabbing him up, I backed slowly out of the duck pen, my eyes trained on the two Nasties should they decide to refresh their attack…. Mr. Peckerhead, his eyes filled with shame that these two tough broads got the best of him. He had tail feathers missing, neck feathers missing, and one eye was beginning to swell shut. As I carried him back to his coop, I couldn't help but feel sorry for the little Bad Boy. And I couldn't help but chuckle as to how fast the Little Nasties had taken the wind from his over inflated ego… and handed him his ass in return… and the lesson learned here I think is that no matter how bad ass you think you are… there is always someone out there that is bigger and badder than you. And my Peckerhead learned this lesson by having his ass handed to him on a platter by two very pretty and normally docile duckies.

Stepping Back in Time

So today I decided to give myself a challenge here on the homestead. As if things aren't interesting enough around here already, right? Seeing the news report of Hawaii having their Emergency Alert activated (although it was human error), the panic that ensued was scary. And I wondered, how would I handle an emergency situation such as that? Because you know, if in fact it did ever happen, life would change as we all know it. No power, no running water, no heat for those who are on the grid, grocery stores shut down with empty shelves even if they did survive the blast…. So how would I manage? Yes, I have stocks of food (canned veggies, stews, beans, meats, fruits) so food wise I would be fine for probably about a month or so. But no power? So today I shut down the power to my house (other than my fridge)… and decided to go cold turkey….

Well let me tell you! It ain't as easy as one would think! My morning chores with the Little Chicken Mafia, The Bad Boy of the Barnyard and The Little Nasties were almost impossible… not quite.... But you get my drift. Carrying my lantern, I went out to the well and about gave myself a hernia trying to slide the cement cover off to draw water…. but I did it. After attaching a rope to my black bucket, I pulled up a five-gallon pail. This was just to water the beasty babes of the prison yard. Carrying the bucket in one hand, the lantern in the other, I made my way through the dark and into the coop. Last night we had a light dusting of snow… and it covered the icy patches on the walkway…. So, I proceeded to slip and slide all the way. Inside the coop the ravenous little dinosaurs greeted me with squawks, clucks, quacks and of course, a growl from Peckerhead. Dim light, dinosaurs crowding my feet… me tripping and water bucket flying to empty over yonder on the far wall… fudge me!

Another trip to the well, another pull of five gallons to water the animals. My arms were already feeling the burn. But I accomplished my goal, and the beasties were happy. So back into the house where it is still pitched black. Setting the lantern on the table, I stoked up the wood stove to make coffee. Went to turn on the faucet… oops, no water, so huffing and grumbling I grabbed the lantern and went back out to the well… another five-gallon bucket of water. Setting it on the floor my dog decided she needed a drink and before I knew it, her head was submerged in water that I had planned to use for cooking and drinking… it took everything in me not to boot her in her black little butt…. Placing a big pot on the woodstove I started heating the water for dishes instead. Back out to the well for more water, this is four trips to the well already. Okay, I got this.

Water boiling on the woodstove, coffee made, and I am re-thinking this no electricity thingy. And I was realizing just how much I was dependent on the grid… on the conveniences of electricity, running water, TV, internet, refrigeration. As I sipped my coffee, I realized breakfast without my electric stove and microwave would have to be planned out….

I've cooked on a wood stove many times in the past but never breakfast, mainly stews, soups and roasts.... So, getting out my cast iron pan, I let it heat up on the woodstove, re-stoked the fire and set another small pan of water to boil for oatmeal. Normally cooking breakfast takes me all of ten minutes… but not this morning. Working by lamp light I cooked an omelet, got the oatmeal to boiling and set a slice of bread beside the pans on foil to warm it up. So far so good…. One half hour later I was eating. I still had to think of lunch and supper… because honestly, cooking on a wood stove is not as fast as cooking on a regular gas or electric….

Using heated water (the dog drank out of) I filled one dish pan with soapy warm water and another dish pan with clear water for rinsing. Okay, so after these morning dishes any others I use today will be stacked and washed tonight… otherwise, I am looking at more trips to the well, more water being boiled on the woodstove. And although it's only a little after breakfast I am already pooped out.

Thinking I would have a leisurely few hours to relax, I realized that if supper was to be ready by this evening, I would have to start now on preparing it. So off I went to the kitchen and by lamp light, began to peel and chop vegetables and potatoes for the stew I had planned for dinner. Browning the beef cubes on the woodstove in my cast-iron skillet, again having to go to the well for more water to cook the stew, finally I got everything into the stew pot and simmering on the stove… the hell with making biscuits as I had planned… and hubby best not complain one iota that there is no bread for dinner… or he just might end up with my soup pot on his head.

It is not noon yet and I am already exhausted… and I wonder in amazement at how those pioneer women ever handled these seemingly simple tasks day after day, year after year. Now I pride myself on being one tough lady, but I will tell ya, those pioneer women had it all over me…. And we have nerve to bitch and complain about how hard our lives are… one morning without the convenience of electricity or running water and I'm ready to cry uncle….

Happy Anniversary

Well, in light of my 39th Anniversary yesterday I feel it would be remiss of me not to write an Ode to my husband. And really ladies and gents, just to peck about him. As you all have learned, we live here a simple life on this homestead… and there are times when he and I just have to get right down to it when tackling a problem. And our problem on this day was the wood stove. Not really a problem but more of one of those maintenance duties that neither of us like to have to do. Now before I begin, for all you gents that may be reading this, ladies too, I am not hubby or man bashing… although there are times when I seriously have thought of bashing hubby with a two by four for pissing me off. Why? Because my husband has the patience of a toddler! And when we have to work together on a project, the two-year-old in him comes roaring its bratty little head to the front and center. And in turn, this turns me into a complete and utter mad woman! He does know how to push my buttons, no surprise after thirty-nine years together.

I was in the middle of beating down the bread dough when he decided that the woodstove chimney pipe needed cleaning. Mind you, I have my hands covered in dough. I look up.

Me: "Really?"

Him: "Yeah, might as well get it done."

Me: "You don't see that I am busy right?"

Him: "I got this. You don't need to help."

Me: Shaking my head and grinning. "Okay. Whatever."

Now cleaning the chimney pipe is a two-man job. Always has been, always will be. The pipe itself is awkward and heavy, slippery to hold. When taking it down, I will usually grab one end while he grabs the other so that all the creosote from the inside doesn't come spilling out onto the living room carpet. Because that mess is a bitch to clean up. So, with a sinking feeling in the pit of my stomach I watch as he-man wrestles this 10-foot-long pipe out of the ceiling. The words coming from his mouth make me laugh… this is the two years old rearing its bratty head. Wiping my hands on a towel and throwing my dough back into the bowl, I walk over to lend a hand.

Him: "I got this!"

Me: "The fudge you do!"

Him: "You're in my way!"

Me: "Shut up already! Stop being grumpy. Fine! Do it yourself a*** hole!"

Stepping back, I watch as he pulls the pipe slowly from the ceiling… just knowing what is going to happen next… and sure as shit, it did. The pipe came crashing down… filled with black ash and creosote, covering my living room carpet and sending up a cloud of nasty ash dust that covered everything within a three-foot vicinity including my bread dough.

Me: "Nice going dickwad!"

Him: "It slipped out of my grip."

Me: "Well? If you'd have just let me help, not

been so impatient, we might have gotten through this with minimal damage… now you can clean up this fudging mess!"

Him: "Don't worry, I will! After I get this pipe cleaned!"

Now I know his idea of cleaning and my idea of cleaning are about as far off as the north and South Pole. And my bread, well that is a total loss. And my temper, I can feel it starting to rise like the sun on a hot summer morning.

Me: "Ya know what? Don't fudging worry about it… I'll take care of it!" At this point he can hear the edginess in my voice.

Me: "You are so fudging pig headed Gosh, I really, really, really dislike you right now!" My temper is winding up and I can feel a bad one coming on.

Him: "Honey, get a grip… it's just a little ash ... No biggie…."

Me: Frothing at the mouth and growling like a banshee… glancing on the table to see if any weapons are handy and within reach.

Him: "Help me carry this outside, will ya?"

Angrily I grab the other end of the stove pipe which has already dumped its contents onto my floor… and leading the way, we take it outside to finish douching it out. I can hear him behind me chuckling.

"What the fudge is so funny?" I snap over my shoulder.

Him: "You look a mess… you've got chunks of ash in your hair."

Me: "I so hate you right now!"

Throwing down my end of the pipe I left him on his own….D.I.V.O.R.C.E!!!!!!

Yes, my friends… thirty nine long and happy

years!

The Illustrious Mouse

I woke to a throbbing and painful toe.

I woke with the remnants of yesterday's headache still lingering.

I woke to a salmon tinged sky, a few dark clouds but the promise of a warmer day.

I woke to the sound of the prisoners rattling tin cups against the bars of their coop demanding breakfast. Would I be facing a riot this morning? We'll see.

Let's start with the throbbing toe... and the headache.

Yesterday I woke with a headache… just a slight one but enough to cause me to lie down mid-morning, close the curtains and close my eyes in hopes it would go away.

Resting lightly, I could hear all the little creaks and groans that are normal for this log cabin. I dozed, drifting off into twilight; my headache sliding back gently and my body sighing into the couch. The dogs, Sadie, Lucy, and Tara-belle, were quietly dozing on the floor. Just as my body and mind started to slip into that Zen state of dozing, I heard it.

The gnawing and chewing of our resident mouse. Jumping up off of the couch, I flew to the kitchen sink and opened the cupboards… not thinking I'd catch the fuzzy little mischief maker but rather hoping to scare it enough to get it to stop eating my house. With three dogs crowding me, I moved bottles of cleaning supplies, a mound of bags, and various other items we collect under the sink.

That is when all hell broke loose in my kitchen… dogs, the cat, the mouse, and me in the middle. For whatever reason, this fuzzy little critter decided to run out of the cabinet and right onto my stockinged foot… I screamed, the dogs (all three of them) pounced and the cat let out a howl that would rival any mountain lion on this planet.

Sadie was the first one to bite at the mouse… which was now frozen in fear on top of my foot… she missed the mouse but not my foot. Yelling, I shoved her away and kicked the mouse from its death grip on my foot… it went flying against the cabinet door and came to rest on the floor in front of me.

Tara-belle lunged, stepping on my foot, driving her nail deep into my toe… the ring toe… next to the pinky toe… and the little piggy will now never make it to market… as it was gushing blood. And this human was screeching. The mouse was skittering… the cat was howling and Lucy, quick as a flash, grabbed the fuzzy mouse in her jaws and thought she was going to have a mouse for breakfast….

I yelled… Lucy dropped the mouse and Tara and Sadie made a mad dash to get it… and I made a mad dash to keep them from getting it… and there is blood now spattered on the kitchen floor… and the mouse… well it lay in a fuzzy little heap, curled in on itself…

Beating the dogs off, I checked it for a heartbeat… and felt a tiny little quickening in its chest. I thumped it gently to see if it would move. I nudged it…. poked it… to no avail. It was alive but frozen in terror. Grabbing my oven mitts… I gently picked it up… meanwhile I have three dogs crowding my legs, jumping on me and snapping their jaws…. I was not impressed with their rude behavior and told them so… sending each of them to their corners….

Leaving bloody footprints across the floor, I carried this poor fuzzy little dude out to the woodshed and placed his/her little body on top of a piece of wood. My hubby won't be impressed…. but hey, I cannot just kill this poor little dude/dudette….

So that is why my toe is throbbing this morning… and my first chore this morning will be to take a walk out to the woodshed and see if this fuzzy little mischief maker has made it through the night.

Sunday, Sunday

I love Sundays on the homestead. For me it is a time to contemplate the week behind me, gather my stories and spend the day with nurturing projects in my kitchen. So, after feeding the hordes of hungries, scooping the poop from the doggies and walking the poop patrol in the door yard, I gravitate toward the kitchen to start my daily doings. And today on the agenda is bread and medicine….

By six a.m. I've got my dough mixed, beat, and rising. Three rises then it will be ready to bake and the warm smells in my kitchen will make my mouth water… there is nothing on this green earth better than warm bread and real butter….

As this goodness is going through its first rise, I pull down a variety of herbs I will be working with today. I am making Transition Time Oxymel… a delicious blend of dandelion root, nettle and astragalus root… mixed with apple cider vinegar and later, honey which will end up being a sweet daily addition to help with digestion and immunity…. I love that I can create these medicines for my family and friends. Secondly, I am making a Lemon-Rose Bite and Sting Soother, a blend of lemon balm, rose petals, calendula flowers and witch hazel… this will come in handy come spring when the pterodactyl's and buzzing bees bite….

Lastly there will be tinctures steeping in vodka, catnip, chamomile, lavender, eleuthero and astragalus, all providing nourishing medicine for the various ailments of the body….

It amazes me that I am gifted this knowledge.... It amazes me that I can look to the earth for what I need to keep my family healthy. I think in these times of convenience we tend to forget how fulfilling these simple tasks can be. Baking bread, making medicine, caring for our animals… all nourishing activities that balance the mind, spirit and body. Living simply is not always easy. Visit me on a frigid winter morning, where my breath puffs out in frozen mist, and you'd see toes and fingers that are cold, boots that are wet as I tromp through the snow for my morning breakfast and to slog through the daily chores. But ask me on a Sunday such as this if I regret one moment of my lifestyle here on the hill and I'd give you a big HECK NO! I am doing and living the life I love, hardships, struggles, joys and laughter… and making medicine and bread….

Fun Days

There are days on the homestead that are just magical… delightful in their own way and today happened to be one of those days. Hubby and I had been having trouble with our woodstove, and try as we might, it was one of those jobs that we just couldn't do alone. So, in answer to our struggles... we called our son who is not afraid to climb on the roof and dig in with helping his dad. To my surprise and joy, he brought my two granddaughters over to visit while he worked with his dad. Now Cora, the two-year-old and Charlie, the five-year-old are at that age when everything is interesting and fun and Oh, so magical. And I love sharing in their magic. And they are both becoming quite the homestead children. So, while the men worked, we spent time looking for the chicken eggs that I hadn't yet gathered… and their happy squeals of joy and laughter filled the coop as they dug their little hands into the hay in search of the hidden treasure. Petting the chickens, talking in all manner about poop and bugs and mice, we forayed deep into the world of my little flock. Each of them filled their pockets with their treasured eggs and happily ran them into the house to put in my fridge.

Next was sliding on the big hill in the back yard… a hill to them that looked, as Charlie squealed to me, ginormous! Although in my adult eyes, it truly wasn't that large. Laughter rang in the air as time after time the two girls carried the sled up the hill and loaded on to have Grammy give them a big push to slide back down. With the warm temps and the sun shining, with the laughter and the fun, my heart sang as only a grandmother's heart could. We were making memories that I hope will last the girls a lifetime. After an hour or so of sliding I saw that little Cora was getting a bit chilly and tired, so I bundled her up into my arms and took her in by the wood stove to warm. Charlie, energetic and curious, balked at the idea of leaving the fun so I gave her a special chore to do while I was inside with her sister. I asked her to gather up the flock which were free ranging on this sunny day and scoot them back into the coop. Really, I didn't think she would be able to, but I figured this would keep her busy just for a bit. Well to my surprise my little homesteader not only gathered up the rowdy little chickens, she shooed the Rooster in as well! Being only five, she couldn't reach the lock on the coop door so in order to keep the flock from wandering back out, she grabbed the biggest rock she could and placed it against the coop door….

When I asked her proudly how she had managed to gather up all the chickens, she replied with a grin: "Grammy, I did just like you do! I spread my arms wide, flapped them and shooed the little ones right into their coop!" I couldn't help but laugh. I hadn't thought about how much this little girl pays attention to me when I am doing my chores, but I will tell you all, she certainly does… and I am so proud at all my granddaughter has learned about my life here on the homestead.

As a reward for helping with the chickens, we decided a walk deep into the woods would end our day of fun. So, hand in hand we went exploring and Miss Charlie shared with me her delights and dreams, her magic and wonder… we peeked into fairy holes in the trees and she swore to me she saw wings… and I have no doubt she did. We found our special place that we picked out over the summer where we made plans to build our secret shelter with sticks and branches from the surrounding woods. We found our sitting tree where we spent many a summer afternoon sitting and listening to the forest sounds around us. We tentatively crossed the swamp where we giggled and squealed all the while hoping we wouldn't break through the ice and fall in… me knowing it was only a few inches deep but her believing we were going to sink up to our necks… and this brought squeals of fear and excitement as only a little girl experiences….

It is these moments that make me fall in love all over again with my life here on the homestead where I can share the joy and the passion, pass on to the little ones all the wonders and magic of living so simply and so close to the land. Although my grand-girls will grow up and may decide that this kind of lifestyle is not for them I know they will carry with them the magical memories created here on Grammy's homestead… and perhaps, who knows, may decide one day that this magical life is something they would like to try for themselves. I can always hope right?

The Lockdown

The trees are covered with ice and it is a letdown from the beautiful warm temps we had over the weekend. Laundry has been started and the big hand on the clock above the fridge is just reaching five a.m. My day has started. My mouse is alive and well out in the woodshed, she/he peeped its furry little head out at me yesterday. I listen as Peckerhead crows out his morning call…. I am mad at that fluffy little flucker… as I limp around the house doing my chores like some old lady with arthritis in her joints….

No, I don't have arthritis. I am sporting a bruise about the size of a saucer on my knee which also has swelling the size of an orange. And it is from chasing an escaped Peckerhead convict around the icy yard yesterday, slipping and falling several times, as the Little Chicken Mafia cheered him on. They wanted him to embrace his freedom…. I wanted him back into the coop. The Little Nasties quacked behind their fence… wanting and hoping for another go at this handsome dude. They were preparing for a rumble; I was preparing ideas for a good stew. Needless to say, I won the race, and he went back into lockdown. He was not a happy prisoner… crowing out his displeasure as I closed the jail house door.

So today I will go into the coop armed with a broom. Why? Because Peckerhead has been known to hold a grudge… and I want to be prepared in case he decides to strike. I know him and the girls, the rowdy, conniving little bunch they are, have been plotting and planning my demise. It is not a matter of if but rather when! And I will be prepared. No more prison breaks on my watch. No more sneak attacks if I can help it. Awareness is the word of the day… be aware of my surroundings… if anything looks suspicious… take notice and take action…. I feel like I'm preparing for my own little Chicken/Rooster Terrorist attack. It is an uneasy feeling.

Now, I have had people who read my stories ask me why I would keep such a beast around, and I always tell them it's a complicated subject. But the basic bottom line is that I've come to respect this little dude. He is a great protector for the Little Chicken Mafia dudettes… he'd lay his life down in a Mississippi minute should a fox, coyote, hawk, or other predator come after them. He even protects them from me. He's got spunk and attitude, something I find challenging and endearing. He's got a big and bold personality!!! And as wily as he can be, he just has a way of looking at me that makes me smile. Yes, we have a very dysfunctional relationship, he and I.

So today I am armed with a broom… and an attitude of my own. And I am sure he's already got his moves all planned out… and we will do this dance one more time.

Midnight Madness

Well I woke up tired this morning so when the alarm went off at 4:45, needless to say, I just wanted to snuggle back down under the covers and get a few more minutes of shut eye. I could hear Peckerhead calling from the coop and I could hear hubby moving around in the kitchen.

Last night was another night of the MOUSE! Dogs, cat, and mice do not make for a happy me.

Hubby: Getting out of bed to feed the woodstove. Then into the bathroom. Now when I get up at night, I am always considerate and quiet. Not my hubby and I swear he does it purposely, making enough noise to wake the dead.

"Tara! Tara!" I heard as he, for whatever reason, was yelling at the dog. Next, bang, bang, bang! As Kitty girl rattled the cabinet doors.

Me: Rolling over tiredly. "What is going on out there?"

Hubby: "Tara!" Bang, bang, bang!

Me: Now getting really pucked off! "Why ARE you yelling at the dog! Please put the fudging chair in front of the cabinet so Kitty girl will stop banging!" I hear him slide the chair across the floor. Now, not wanting to miss out on the fun, my two little girls decide to join in on the chorus with high pitched whimpering and whining. The top of my head is about to pop!

Me: "Sadie! Lucy! Knock it off!" I screech.

Hubby getting back into bed.

Me: "What were you fudging at Tara for?"

Him: "Because the cat is under the sink trying to get that stupid mouse and Tara was rattling the cabinet doors."

Me: "Did you put the chair in front of the doors?"

Him: "Yeah, I think Kitty girl is still under there though."

Me: "So, you trapped the cat under the sink? Really? Seriously? What the Fudge is the matter with you!" I yell as I jump out of bed.

Him: "Well you told me to put the chair in front of the ***** cabinet!"

Me: "And since when do you EVER listen to me, moron!"

Him: chuckling. "I always listen to you! Everything you say is SOOOOO important to me."

So, at this point I am seeing red…. I can't catch

my breath, anger is pulsing like a drum beat in my ears and I have visions of beating my husband to death… better yet, I entertain the fantasy of letting him fall asleep and then stuffing a pillow over his face… the words 'Till death do us part' take on a new meaning for me.

I swear to all that is holy I will one day snap and lose it on this man! Stomping out to the kitchen I drop kicked the chair out of the way and sent it flying across the floor in a fit of temper… I whip open the cabinet doors and spend five minutes trying to catch Kitty girl who is moving from one end of the cabinet to the other staying just out of my reach. Finally grabbing her by the scruff of her neck I pull her out of the cabinet and hear the low growl as she has clenched in her jaws a little field mouse… it was then that Tara-belle decides she is going to climb up my back, all sixty pounds of her, and the cat decides to unleash her claws at Tara-belle and I catch a face full of feline fury. Quick reflexes? An instinct to survive? I don't know but before I even realized what I was doing, I had flung the hissing, clawing cat across the room (I do really feel bad about doing this)…. She landed with a thump on all fours with mouse still securely in her mouth. Until Tara-belle pounced! Then the rumble was on! Chairs knocked over as she chased the cat, me screeching as I chased the dog, kitchen table crashed into spilling the sugar bowl and shattering it… hubby yelling and me trying desperately to get my hands on one or the other so I could stop this midnight madness erupting in my kitchen….

So, this morning I sit here and sip my coffee… hubby is talking to me and I can only respond in disinterested and grumpy grunts and growls and dark, menacing glares. I know later today I will be having to search for the headless mouse somewhere upstairs because unfortunately the little pooping fuzz ball didn't make it…. Tara-belle is sleeping quietly by the woodstove, the cat is purring on the sofa… and I am plotting and planning of how I can interrupt their sleep as they did mine….

And lastly, a nap is on my agenda for later today.

No Respect

Let's just say, as a homesteader, I have learned many things… and one of those things is that animals do not respect personal space or boundaries. And if you ever plan on becoming a homesteader… all bets are off as to privacy… in any way, shape or form. Now with my girls (the dogs and cat) we've discussed this issue of personal space many times. And others who may be within hearing distance will tell you the phrases most often said in my house are:

"No, your nose doesn't belong there!"

"Do you mind? A little privacy here, okay?"

"What is your nose doing there? Get your nose out of xyz's butt!"

Now anyone with dogs or cats can relate, right? Going into the bathroom, I have to race the girls (the dogs) to the door and shut it tightly before they can wiggle past me… three faces watching while I am doing my business is neither appropriate nor acceptable. Taking a shower… I have black, wet noses stuck against the glass shower door and on the off chance my hubby tries to be affectionate… well.... We won't go there.

But the incidences of the last two days have shown me how little respect there is on this homestead for my personal space. One happened yesterday when two of my hens, Gladys and Daisy, decided to make their way under the coop and nest.

As I had let all the prisoners loose for a little exercise outside of the pen, I kept a close eye on them. This time of year, there are lots of hungry predators in the area and chicken dinner is an easy meal. Well, I turned my back for just a moment and these two girls disappeared and I heard Peckerhead calling anxiously. So hastily I set off looking for them. Finding them nowhere in the yard I pulled out a cupful of food and placed it in a foil pan, rattling it… and sure enough I heard clucks from under the shed. Now under the coop is not the ideal place to be… as I have a skunk that frequents this area and a possum that I've seen in and out of there… so my chore was to coax these two little prisoners out. I will tell you it's easier said than done.

Getting down on my hands and knees on the melting ice and mud (think downward dog yoga pose) I peered under there and saw them both. I softly coaxed and pleaded for them to come out… I bribed them with treats… I hissed at them… I poked them with a stick… which only drove them further away and under….

As I am struggling to get these fluffy, feathery little poop machines out, I feel a large head come between my legs… Tara-belle has decided to join in, curious to see what mama is doing…. and I hear a snicker from the front porch as my husband is laughing while watching this sixty five pound dog trying to crawl under me and me struggling to push her away… to make matters worse, I had forgotten about Peckerhead… and forgotten that you never, ever, ever turn your back on a rooster….

Covered in mud, freezing from kneeling on the ice, swearing at the dog and the hens under the coop who were now squawking and clucking wildly (they were telling me off in no uncertain terms… the language!!!!) I now feel a pinch on the back of my butt… then another and another… and roars of laughter coming from the porch as my husband watches my rooster peck my arse….

Now I want to say he was protecting his girls… but I honestly can't! This little bad boy was taking his best shot! This bad boy was finally believing he had the upper hand (or beak) in this case! It was payback time, and he was going to be sure I got what was coming to me…. Flipping over on my back I kicked out at this menacing little dinosaur with wings… and booted him clear across the yard where he squawked and growled at me angrily… glaring with his beady little eyes. In a fit of temper, I shouted at hubby who was laughing so hard I thought he was going to have a stroke… I yelled at Tara-belle who went running to hide behind hubby and I yelled at the hens under the coop telling them I hope they became dinner for some hungry creature….

So needless to say, today I am sporting several tiny bruises on the derriere….

I was cleaning the Little Nasties sleeping area. The weather was beautiful, and it was a good day to tackle this nasty little job. So, shooing all the feathered creatures out of the coop, I pulled down the wire separating the Little Nasties from the Little Chicken Mafia. Down on my hands and knees… (Yes again), I began pulling out poop clotted hay and shavings. Gagging and trying not to blow my breakfast all over the floor, I loaded armful after armful into the wheelbarrow. Grab an armful of soggy, poopy, heavy hay, stand up and take it to the wheelbarrow, kneel down and do it again. After about the third trip down onto my hands and knees I turned my head and see that I am surrounded by hens… between my legs, behind my back, over my head on the roosting pole… hens watching me with beady, curious little eyes… the one hen I didn't see was in fact the one I guess I should have paid the most attention to….

Turning back to my work, I suddenly felt claws on my back… then I felt my hair being yanked…. in my haste (thinking it was Peckerhead making another attack) I stood quickly and knocked my noggin on the roosting pole… and flailing my arms I did a merry, disjointed dance trying to rid myself of this obnoxious little flucker who was now trying to snatch me bald… all the while the Little Chicken Mafia squawking and yes, laughing at my predicament… and in my wild twisting and turning I saw Peckerhead standing and staring at me with malicious glee in his little eyes….

One of the hens had decided to make a go at me and I just know he put her up to it… if he couldn't get the best of me... he was going to try another tactic… because he knows I would never expect an attack from one of the hens….

Now, I can honestly say I don't think she was trying to attack me… or my hair. I like to think that she just wanted to sit on my shoulder and curiously watch what I was doing... and my hair... Well; I think she was pulling it for her nest….

But another part of me knows… knows that Peckerhead is plotting and planning my demise and if he has to get the girls in on it, well by God he will.

Yup, that's my story and I'm sticking to it! And as I walked away, I turned and looked at him and smiled.

Take that you little flucker… I win again!!!!

Reigning Queen

The super Blood Moon. I could wax poetic about it and as I gazed upon it last night, I can say that it was absolutely beautiful… but did you know….

Now it may be an old wives tale… or just a farmers tale… but animals do react to the moon… and (some people too… try working in a nursing home or a prison on the night of a full moon)… and on this homestead, my animals… specifically the prisoners, could feel the shift. Peckerhead crowed late into the night… so much so that I suspected a predator and ran out to the coop several times under the light of this beautiful moon to check. Even the Little Nasties were more vocal than usual. After several false alarms… I attributed his rowdy crowing to the fact that he just wanted to be annoying.

So this morning when I went out to the coop

for my usual routine, poop patrol, collecting the eggs, feeding and watering the prisoners… it surprised me to see Peckerhead greeting me face to face as I opened the door… backing up a step or two, I eyed him warily as he curiously peered at me.

"Okay big boy…. just back up a bit so I can come in." I coaxed softly. It was too early to get into a rumble with him and to be honest, I just didn't have the energy.

Him: "Not a chance… go by if you dare…." He growled…. although he didn't actually say this in so many words… (Really, I know my rooster can't talk)… his body language told me everything I needed to know.

Me: "Okay asshat… I will move you if I have to!" I threatened… now growling too.

Him: "Try it and you'll lose an eye…" he warned… ruffling the feathers on his neck which tells me he's going to stand his ground. The Little Chicken Mafia are now beginning to squawk and grumble… nervously edging him on. I look at them and hiss… the Little Nasties are quacking furiously wanting only to get out of their coop and get outside. I pray to the God of Homesteading… a short, but incredulous rant: "Good God, why? Why is it every morning I have to fight with this cantankerous rooster? Just once, just once can't you lay a godly hand over this asshat and make him play nice?"

But Peckerhead stands his ground barring the door… and to get by him is going to mean mixing it up. Reaching out I try to push him away and he growls and lunges at my hand, narrowly missing… he wants blood. Backing up, I look anxiously for a stick to poke him with… I can hear the noise of Welcome to the Jungle playing in my head…. screw you Guns and Roses!

Tucking my head like a football player I throw myself through the door of the coop, scattering the Little Chicken Mafia, wings fluttering wildly, feathers flying, clucks, quacks, growls and hisses, screams of outrage and grunts of anger and I grab that fluffy little flucker by his legs… thank you God I am wearing a heavy jacket because the fight that ensued would have left me bloody and whining….

He reared his head, stretched his neck and pecked at my arms and hands… he kicked those pterodactyl drumsticks and clawed… he hissed and flapped and twisted and turned… and I held on for dear life, because I knew, I knew should I let go he'd have kicked my 120 pound arse….

Now, the thing with roosters is just this, if you hold them upside down by their legs they will, for a few minutes, struggle to beat all hell… but then they will go quiet… and they will be eerily still, almost like they are dead… but they are not. What is happening is all the blood is rushing to their small dinosaur brains… and effectively almost putting them into a Zen state… and I was determined he was going to enter this Zen state if I had to kill him to get him there!

When I finally set Peckerhead down after twenty or so minutes of carrying him upside down, I saw an expression of hatred and shame on his little face... and acceptance for the time being that I am still the reigning Queen of this little establishment… will he challenge me again? Of that I have no doubt… but it won't be today.

The Other Girl

Today I woke up earlier than usual… we have a storm rolling in and things to be done before it hit… but somewhere after the Little Chicken Mafia, The Little Nasties and Peckerhead were fed, watered, and given extra hay for bedding, I wandered off of the routine tasks of this homestead and ventured into the world of my writing. And this is okay….

I am with a writing group… and today's assignment was to write a story with the title, The Other Girl. So, after thinking about this for a while… I have to share with you my thoughts… and a coffee if you are so inclined to grab one. The Other Girl is me… through and through, and the Other Girl is many of my sisters who follow this same homesteading path… the ones who choose to live this life, play in the dirt and the poop, raise their animals, sweat under the hot summer sun and freeze in the cold and icy winters. We are the ones who trade factory for farming, nursing for non-stop poop, desks for damning when the black flies of spring are eating us alive.

In short, I believe I've always been the Other Girl. The one who'd rather be digging in the dirt, canning in the kitchen and slopping my way through mud and poop and attacking pterodactyls…. The one who runs from mice, swats at bees, takes cover under the kitchen table from bats flying overhead. The one who crawls under the house amidst the mice poop, dead carcasses and bugs to fix a plumbing issue. I am the woman who traded in high heels for work boots… a skirt for a ratty pair of jeans and perfectly manicured nails for work gloves and blistered hands.

My life revolves around chickens and roosters,

ducks and dogs… gardens and herbs. My favorite spot in the house is the kitchen where I cook meals from scratch, make medicine, can for the seasons, and bake breads. Where I sit with my grand-daughters and teach them these things I know and hope one day, they too will love this lifestyle I live. Over the kitchen table I share impromptu breakfasts with my grown daughter, share laughter, tears and fears…. And I share quiet suppers with hubby after a long day of work.

So yes, here's to the Other Girls… the ones who share this passion of living simply, loving passionately and thanking God each and every day that we are who we are.

Unsuspected Surprise

Woke this morning almost too tired to put my feet on the floor… but with this storm of a few days ago, my Little Prisoners are chomping at the gate to get out from behind their bars and get some exercise in the pen… and that means, shoveling. The frosty air met me as I set out before daylight to perform my usual poop patrol… and I could hear the clucks and quacks as well as the crows from Peckerhead urging me to hurry the hell up.

Entering the coop, I was met at the door by none other than the Little Nasties who had escaped their pen… looking around I spied Peckerhead up on the roost, far away from these two little terrorists. I could see he did not want a repeat of getting his arse kicked by these two brutes. He glared at me and I could read in his expression how angry he was that they had succeeded in their jail break.

"Don't worry, I'll get them back into their pen." I sighed tiredly. Peckerhead hissed.

"Alright already!" I snapped. Honestly, when I am this tired, arguing with my cantankerous rooster does not bode well. The Little Chicken Mafia are gathering around my feet and legs, tripping me up as I pour feed into their dish… and they attack the dish and my hands like ferocious hungry little fluffers… patiently I shoo them away…. Peckerhead is still on the roost… growling and hissing.

"Well, come down and eat." I yell at him. I know better than to reach up and try to lift him down. He eyes me defiantly. I think he's planning a hunger strike. Gathering up the Little Nasties I scoot them back into their pen and fix the door, feed and water the little boogers amidst the angry quacks… they were enjoying their freedom. Thinking I was going to escape these morning duties with little drama, I carelessly let down my guard….

And….

This is when the poop hit the fan!

Reaching into the hay pile, I disrupted a mouse. Now this wasn't your normal mouse, this mouse was fat… this mouse was a monster…. this mouse ran across my hand, up my arm, over my shoulder… and Peckerhead saw it… with a hungry, predatory gleam in his beady eyes… he zeroed in while I was dancing the heebie jeebies trying desperately to dislodge this monster mouse before the fuzzy, fat little beast got tangled up in my hair… all the while screaming prayers to the God of homesteading. It sounded something like this:

"OH, DEAR GOD! WHAT THE FUDGE… NO… GET THE FUDGE OFF OF ME… GOD WHY? WHY?"

Roosting pole above me… forehead ramming into it… dislodging Peckerhead from the pole who lands claws first onto my chest… beating, flapping, fluffy wings… squawks… arms flailing… Little Chicken Mafia laughing and scattering as I battle the beasty mouse and the cantankerous rooster…

The mouse hit the wall, the rooster hit the wall… and I ran! Right out the door and went ass over tea kettle over Tara-belle who was waiting patiently for me outside of the coop….

Sitting in the snow, I was left stunned by what had just happened. Tara-belle licking my face, my arse freezing, listening to the commotion of Peckerhead chasing the beastly mouse through the coop….

And today I will be setting traps… with lots of peanut butter and cheese….

Lockdown

Yesterday on the homestead was a tough one… a nasty little virus snuck its way into this girl and kicked my butt seven ways to Sunday…. So after zipping out to do my normal morning routine… and zipping is probably not a good word of choice, rather slinking my sick and tired self out the door in my Eeyore bathrobe, my pink muck boots and my barn jacket, I met with the prisoners and explained the situation to them… and surprisingly enough they understood…. even Peckerhead didn't act up with his usual antics….

Four doses of Elderberry syrup, and some happy tummy tea kept me on the couch for most of the day, catnip and chamomile along with lemon balm left me drowsy enough to rest. The good part about homesteading is that in the fall I put up lots of soups… and home-made canned vegetable soup was an easy dinner… especially for those times I don't feel good.

So today is catch up day… set out early this morning feeling much better… and the prisoners met me before daylight clambering for their food and water… and Peckerhead up on his roost like the king of his domain. Now, I must tell you, roosters are well known for being amorous with their girls… and my roo likes one girl in particular… so much so that he has literally pulled all the feathers from her poor back, leaving her bald. Every day I began noticing more and more feathers missing from this poor girl and every day he and I would discuss this situation….

"Peckerhead, enough is enough. You need to leave her alone or you will go into solitary confinement!"

And every day he would ignore my threats… challenging me with a look that said, I dare you! Well today is the day, and I set to do the deed….

After finding a kerchief to cover the poor little missus bald back, I tore down the separation divider between the Little Chicken Mafia and the Little Nasties. I would be moving the dastardly duckies in with the girls and Peckerhead into their little enclosure… Peckerhead watched me with his beady little pterodactyl eyes… hissing and growling now and then to let me know he was onto me.

"I warned you! Look at her! Her freaking back is bare.... It's a wonder she hasn't frozen to death…." I muttered as I kept a close eye on the boy….

In response he hissed at me. I could see the fight coming on… and his pissy, quick temper rising.

"Nope, you've done it! You are going to be in solitary till her feathers come back!" I yelled at him. Meanwhile he's pacing back and forth on the roosting pole… hissing and growling… plotting and planning as to how he is going to kill me. I flapped my gloved hand at him to let him know he didn't scare me one bit.

Once I got his little confinement area ready, I turned and gave him the look. I knew I'd have to grab him quickly or the chase would be on. And although I was feeling better, I still wasn't up to 100%... in a soft, coaxing voice I approached him slowly, him eyeing me suspiciously… the Little Chicken Mafia watching with little squawks and clucks….

In my haste and focus I had forgotten to shut the coop door…. my own foolishness… and just as I reached for the Bad Boy of the Barnyard, he gave a warning growl, spread his wings and jumped… about decapitating me on his way out the coop door… and into the yard… which still has a foot or more of snow… me hot on his arse….

For over an hour I chase that fluffy little flucker through the snow, slipping on the ice, around and around the coop, through the woods, over hill and dale… until I was drenched with rain, sweat, snow, hay, bits of sticks in my hair… and ready to give up… for the first time…. I was beaten… I was humbled… I was crying UNCLE… as this fat little flucker kept a few feet ahead of me at every turn.

It is now almost noon time and I can hear Peckerhead's calls of triumph echoing through the yard.... I can hear his taunting... as he wanders unbridled... free... happy and Oh, so proud of himself.... I can hear the Little Chicken Mafia's party going on in the hen house as they cluck congratulatory words through the coop door....

I'll give him a few more hours in the rain while I rest before giving chase again... let him revel in his short lived, sweet taste of freedom... then it's into solitary for this fluffy little beast.... I think a thirty-day sentence should do the trick.... Teehee....

Sour Crop

As a homesteader, every once in a while, I get a call from my homesteading friends... usually looking for help with one of their animals. I wasn't surprised when my girlfriend called me with a chicken question.

Me: "Hi."

Her: "Have you ever dealt with Sour Crop?"

Me: "Nope."

Her: "Well my little chicken is not acting right... she's not eating or drinking... and when I picked her up, she puked all down the front of me. I thought it might be impacted crop, but her crop doesn't feel hard.... I'm thinking sour crop."

Me: "Ewwww... that sounds awful."

Her: "Yeah, my coat stinks to high heaven. I brought her in the house and put her into the bathtub to keep an eye on her. I just didn't know if you had any herbal advice to help her."

Me: "No, I don't but let me google it." (As I pull out my computer and google in Sour Crop)

Her: "I read online that if you tip her upside down and rub her crop, it'll help…."

Me: "Mmm mm… I don't know but I guess it's worth a try. Have you tried feeding her water with an eye dropper?"

Her: "Yeah, I put some apple cider vinegar in it… maybe that will help."

Me: "Well, I'd just keep an eye on her for now. Let me try to find some information."

Her: "My coat stinks like hell… gosh what is it with these chickens…." I thought, Yes, get some chickens, it's easy… just throw them some corn every day and they will be fine…. but Oh, no… not that easy at all…. I've dealt with what I thought was an impacted egg... which turned out to be lack of calcium and no egg, I've dealt with predators stealing my favorite hen, and now this….

Me: "Yup, not so easy at all."

So, I hung up the phone and googled…. I saw pictures of chicken anatomy, I read that for sour crop you have to turn them upended and massage the crop, I read that you could feed them small doses of cinnamon, apple cider vinegar, and other such things, I read that you need to make them puke but watch for aspiration which could kill them….

Lord have mercy! Not trusting what I researched, I decided to turn to my chicken groups and put the question out there.

Me: "I have a chicken with what I believe is sour crop. What do I do?"

Reply: "Are you sure it's sour crop and not impacted?"

Me: "Yes, I think so."

Reply 1: "You have to make her puke. Get the sour out of her crop."

Me: "And how do I make a chicken puke? Stick my finger down her beak? LOL!"

Reply 1: "Noooooo… she will bite you… lol." (Insert smiley face here). "You have to tip her upside down, between your legs, massage her crop area in a clockwise circle until she pukes."

Me: "And then I do the hokey pokey and turn myself around right?"

Reply 1: "What?????"

Me: "Joking."

Me: "What if she doesn't puke?"

Reply 1: "She will if there is sour there… if not, it's an impacted crop."

Me: "Okay, what else can I do? I think she pretty sickly…."

Reply 1: "Are you sure it's sour crop?"

Me: "Yes, I do believe so. Her crop is soft and mushy."

Reply 1: "Okay, so once she pukes then you need to keep her warm and quiet. Then you have to make her puke every thirty minutes or so until she empties her crop...."

Me: "Awesome, sounds really gross."

Reply 1: "It is and stinks like hell. But, keep her in away from the others for a day or two, push fluids and you can also give her a bit of olive oil to help…."

Me: "Okay."

Reply 1: "Oh, and she probably will have diarrhea, so you'll get pooped on I'm sure."

Me: "Yay… another shitty thing to look forward to."

So, this is how my girlfriends chicken ended up in her bathtub… and no, I did not suggest to her to do the hokey pokey and turn herself around….

So, what I've learned:

Sour crop = Make them puke by placing chicken between your legs, upside down and massage crop all the while doing the hokey pokey and sending up a chant to the Chicken Gods for rain.

Bathtub = Keep said chicken in the bathtub while feeding her with an eye dropper every half hour, water, apple cider vinegar, olive oil and chicken grit…. thus, no baths for you even though said chicken has puked and pooped all over your legs.

Sleep = very little sleep as you listen to an angry, sick chicken in said bathtub clucking all night.

Having Homesteading Friends = Priceless.

Murder on the Homestead

This morning I feel like a murderer…. I woke to the darkness and took my time out on the deck to enjoy the mornings peaceful and finally, warmer temps. Coffee in hand I could hear Peckerhead crowing his own greeting to what promises to be a nice day. The Little Chicken Mafia clucking away quietly, almost as if whispering. The Little Nasties, a bit louder with their demanding, rowdy quacks. I am coming out of a sleepless night…. I am tired and groggy trying to dig deep to find any little scrap of energy that will carry me through the normal morning chores.

My restless night started at about 1 a.m. Bumping under the kitchen floor, the dogs scrabbling across the room trying to get whatever it was that chose this ungodly hour to make an appearance. Kitchen table chairs scattered and knocked over, Tara-belle scratching at the floor as though she could dig out whatever it was making the noise. Jumping out of bed I sleepily, tiredly made my way into the kitchen where I yelled, I threatened, I stomped my feet all in a fit of exhaustion induced temper. The dogs scattered… Tara-belle to the couch, Lucy under the kitchen table and Sadie running like her ass was on fire back into her kennel… and the critter under my house happily gnawing away on the beams under the floor.

To make matters worse, I heard the mouse trap under the kitchen sink snap….

And… I… don't… want… to… check… it!

You see, I've long come to the conclusion that I run a happy Mouse Hilton up here on the hill. These little pooping pellet machines have left me presents under my sink, on the counter tops, in the pantry, in the bathroom… fuzzy wuzzy has moved in and I do believe he brought all of his relatives along with him…. They are helping themselves to my food, driving my dogs and cat crazy and keeping me awake most nights as I listen to them partying it up in the suite next door…. And rowdy neighbors is not what this homesteading girl needs…. I clean up their mess each and every morning and I wonder, should I give them little mousy beds, put mints out on their little pillows, offer them a fuzzy warm robe?

No… screw that! I instead wage war on these cute, big eared, fuzzy little pooping machines. And tonight, I murdered one. And I feel horrible about it. And it is past midnight…. I don't want to check the trap and reset it… I don't want to drag my arse out into the dark and the cold to toss this now deceased fuzzy little critter into the woods… and I don't want to run into the larger creature that has taken up residency under the floors of Happy Mouse Hilton… I just don't want to….

I go back to bed… and I don't… and I couldn't sleep knowing there was a dead mouse under my kitchen sink… I can envision all his relatives gathering around the mouse trap to pay their last respects… mourning the loss of their dearest fuzzy friend/husband/father…. Little paws clasped in prayer, tiny, fuzzy heads bowed in sadness…. Perhaps I should lay a buffet for them… food is always welcomed at these celebrations of life events…. Right?

And this morning I face the not so nice chore of removing the little fuzzy, pooping machines body… and I feel like I need to dress in PPE (Personal Protective Equipment) to do the job… and I also dread the chore of having to crawl under the house and find out what larger critter has taken up residence there and kick that little blasted beast to the curb….

And in all honesty, I would just rather say screw it all and go back to bed. Such is life….

Trapped

Well, after my mouse and varmint saga of last night, today has proven to be quite the challenge... and there was blood.... Mine.... It started with the Little Nasties raising a ruckus in their exercise pen.... Twice I went out, slip sliding on the ice to see what was the matter... no predators that I could see, food dish full, water dish full, so I was just basically stumped. They were agitated and restless, pacing and quacking.... Because my Little Chicken Mafia have been on an egg strike the past two days, I decided while I was out, I might as well check to see if they pooped me little presents this morning. Upon entering the coop, I counted three red heads... the fourth, Gladys was missing... so searching the coop, searching the exercise pen thinking she made a jail break, I failed to find her.... Meanwhile the Little Nasties are raising the roof with their panicked quacks.... Back inside the coop I asked Peckerhead where his girl was, and he hissed and growled at me menacingly.... Oh, for fudge's sake! Then I heard the faintest little cluck... a whisper almost... from behind the hay bale... then behind the wall....

Gladys had somehow made her way behind the wall of the nesting area and was stuck.... Of course, it was near the top third of the wall, much higher than I could reach... so monkeying up over the nesting boxes, standing on the bracing... I climbed. At five foot two inches, I couldn't quite reach where she was trapped... I could see her poking her little red head up... eyes pleading with me... as I scolded her for getting herself into such a predicament... and she cooed back softly.... Laying myself across the poopy nesting boxes I stretched and reached, her all the while pecking at my hand... thus, the blood... and finally pulled her up out of the hole she was in... thwarting her great escape... as I set her on the floor, I looked down to see my new sweater covered in poop and slime and hay and shavings... the cost of saving one grateful Gladys... and this damn girl better start pooping me some eggs as payback!!!!

A Late-Night Visitor

Last night was one of those nights that just called for me to be outside, staring up at the silky black sky and the millions of bright stars. I relaxed into the night, into the call of an owl somewhere off in the distance and the yips and barks of the coyote's up on the hill. Surrounded by the tranquility of the night creatures as they hunted and roamed.

But, as you know, living this lifestyle you have to accept living with the wildness and I've seen much of that wildness in my own front yard, especially on nights such as this. I've had a bear less than twenty feet away from me, I've had a coyote in the back yard howling, I've had an owl that makes nightly visits, a bobcat that casually walked with in a few feet of me, two raccoons that frequent my compost pile, and possums that love to roam my back yard…. It was no surprise to me when I heard the tin cans (an alert system), rattling on my chicken fence….

Grabbing my flashlight, I peered into the darkness… I had prisoners to protect and I take my job very seriously. I will hug him, I will pet him, I will name him George….

That is the possum who has taken up residence

in my domain. Now possums are curious creatures… usually very gentle… kinda ugly... kinda cute… but they do love them some eggs… especially fresh chicken and ducky eggs… and they have been known to kill a chicken or two… or five…. I have nothing against them. In fact, they are very useful for keeping away ticks and bugs… but… I have an aversion to anything that wants to harm my prisoners….

I watched this furry fat rat like creature climb up the chicken wire enclosure…. I watched him/her climb onto the top of the coop….

So being the warden of this establishment, sort of the chick penitentiary, I could not allow this creature to go any further….

Sliding on my boots and a jacket over my Eeyore bathrobe, I grabbed my big walking stick off of the porch… flashlight in hand I slip/slid across the icy yard and made my way to the offending little beast.

"Scat! Get down off of there! Don't make me shoot you!" I yelled…. Well George just froze, atop the chicken coop, rat-like tail curled around his body, curious black eyes ringed with white fur… and stared at me. I did not want to beat this animal with my stick.

"C'mon… don't make me come after you." I said softly as George, or perhaps she was a Georgette, stared at me with frightened eyes. Fudge! I really don't like violence, but my yelling was not scaring her/him away.

Taking my stick, bracing my feet so they wouldn't slide out from under me, I gave George/Georgette a good whack.

"Get! Get down and run you little fuzzy rat!" I hissed. Poor baby just stared at me and curled into a tighter ball. Man! I really didn't want to open the pen door and get underneath this pitiful creature, but she/he was leaving me no choice. Unlocking the pen, I went inside, all the while keeping my eye on my new-found friend. We hadn't yet established a trusting relationship, and I had visions of this twenty-pound fatso jumping on my head….

Underneath her/him as she was perched upon the wire roof of my chicken coop, I poked upward with my walking stick.

Will you just go? Get! I hissed. Now at this point a prayer to the God of Homesteading was in order… so I set to work praying.

"Really God? Really? As if I don't have enough to worry about with the prisoners you now send me this beast to deal with? Funny… but I ain't laughing… ya hear me God?"

My upward poking with the walking stick wasn't budging the fat fuzzy possum one iota…. she/he was frozen in place… frozen in terror… and by the second I was feeling more and more like a bully….

"Fine! Stay there but don't you dare try to get in my coop you rat like pain in the arse!" Giving up I walked back to the porch and for several hours I watched George/Georgette perched up on the top of my coop… for hours I reveled in the peace and quiet of the night, staring up at the stars, listening to the night creatures….

Ahhhhh the beauty and grace of the homesteader.

Home Sweet Home

What is with all these creatures coming to live at Casa de'Broadley? I mean I understand that living on the homestead I need to accept a few critters but geez! It seems that between the bats, the mice, the moles, the hornets that congregate in the upstairs loft, the possum making her happy home beneath my floors, the ants that come in with the firewood I'd be given a break and at least one creature would stay outside where they belong!!!! But Oh, no… they all want to live with me... up close and personal!

The whole damn family has decided to wage war against me! Mr. Mouse, Miss. Petunia Mouse, all the mousy children, aunts, uncles, brothers and sisters….

And last night was a whole new level of Mouseville Terrorism….

And I lost my shit!

Kitchen chairs scattered, dogs running for cover, all the contents underneath my kitchen sink toss haphazardly across the kitchen floor, Kitty Girl hiding in the upstairs loft… as I chased a mouse… at midnight… through my house… with a broom… and a boot! I was going to squish that little pellet pooper…. I was going to wring its little neck….

But alas…. I did not catch him… the dogs did not catch him… the cat did not catch him, and my house looked like a tornado has just rolled through with all the wrath of hell!

It started just after I got home, about nine thirty… sitting on the couch, minding my own business… reveling in the quiet after a long shift of so much noise, so many demands. A faint scratching right behind my head… and papa mouse sitting prettily on the back of the couch… looking at me. Jumping up, I did the heebie jeebie dance… ewwwwww! And I swatted that little fuzzy, beady eyed pooping machine clean across the room… all the while screaming to high heaven. Hubby yells from the bedroom….

"What the fudge is going on?"

Me: "A mouse just dang near crawled onto my back!"

Him: "And?"

Me: "And what? A freakin mouse… always more mice! Do you not get that?"

Him: "What do you think I can do about it?"

Okay, so here we go... he really wants to go there? I was in the mood for a knockdown drag out. I was tired. I just wanted to eat a snack and go to bed… I was grumpy and my back felt like Hulk Hogan was tap dancing on top of it! And this man snuggled warmly in bed wanted to give me shit? Ohhhh not a good idea… not a good idea at all….

Me: "Burn the house down… I'm sick of the mice, sick of the possums, sick of the flying and biting and crawling creepies!"

Him: "Okay, that makes sense…. Can it wait till morning?"

Me: "Shut up and go back to sleep… moron!"

Him: "Go to bed!"

Okay, so that didn't go all that well…. Nonetheless, I needed to catch the mouse. So, I ran maniacally through the house, knocking over furniture, losing my mind in the madness of finding this fuzzy, pooping little creature… knowing I would not be able to close my eyes until the deed was done….

Into the bathroom I chased him, into the

kitchen, back under the sink, skittering under the couch, back into the bathroom… crashing and banging and yelling and swearing… and he ran, terrified of this big monster (me) who was trying her dangdest to do him bodily harm… terrified that there would be another prayer vigil and funeral procession… little paws clasped in mourning… little fuzzy heads bowed….

I just knew there were many mousy eyes watching us… the epic battle… them rooting for their mouse relative… me rooting for squished mouse so I could sleep….

And sadly, frustratingly the little fudger escaped me… proving he was better at this battle than I would ever be…. That he was the champion of strategic battle planning….

Crawling into bed, I snuggled into hubby.

"Did you get the mouse?"

"Shut up!" I replied as I heard him chuckle….

I eyed the extra pillow, wondering if I could get away with smothering him in his sleep…. tempting… but no. Turning over, I growled….

"Goodnight Asshat!'

Spring

It is amazing to me how great I feel when the sun is shining… perhaps it's because I finally got a good night's sleep with no nocturnal visitors, perhaps it's because I was too tired to care if I did have visitors…. Either way, I woke this morning refreshed and buoyant….

I sat on the front porch, watched the darkness melt from the sky and the morning light bleed in. The sun came up strong… and bright…. and warm against the chill of the early dawn. The Little Chicken Mafia, the Little Nasties and Peckerhead were all calling their early morning greeting… almost as if they knew too it would be a bright and beautiful day. I want spring… I crave spring… and my body, mind and soul melt into this brief reprieve we have from the bitter bite of winter. Stirring through my morning poop patrol, feeding the hungry carnivores, I set to work on some spring cleaning in the house.

By 5 a.m. I had a pot of beans simmering on one burner while potatoes bubbled on another. The bathroom was cleaned from top to bottom, floors swept and washed, trash taken out, laundry folded. Vacuuming was done and bed made. By 8:30 my chores were done. Today I will take advantage of this weather, soak up the sun and clean the coop…. Mounds of thawed poop and soggy hay build-up will go into my compost pile for use later this spring in my garden. With the addition of new kids (goats) coming soon, my mind wanders toward building them a shelter and fencing in an enclosure for the new girls…. I am excited. I will also be getting twenty more little prisoners… chickens to add to my flock… my life on the homestead will certainly get more interesting. I wonder how the prisoners will take to their new roommates.

This is what I've been waiting for all winter, this peek of sunshine, this caress of warmth that stirs the new season of growth for not only myself, but the animals and plants around me. I watch for the trees to bud with anticipation, I watch for the first tender little sprigs of green to pop its head up through the thawing ground with yearning… I see on my travels, lines strung from maple to maple with buckets catching their sweet goodness and maple syrup season begins…. I watch for all the babies that will be born and wander through my front yard…. I listen for the song of the birds as they celebrate another season… and this gladdens my heart and lightens my soul. We have now reached the back side of winter… and what a long one it was… but today I will celebrate the sunshine… it is enough for now.

Crying Uncle

Good morning…. woke up to balmy temps here in New Hampshire… and loving this gift of reprieve from the cold, bitter winter. I have become a mass murderer… and I am beaten to the core… the mice are winning! No matter how many I send to the great beyond, I swear a dozen more come back for revenge….

Pooping little pellets everywhere, eating everything they can get their little mousy paws on… driving my dogs and cat absolutely insane with the nightly chasing and destruction of my house….

Last night I sat out on the porch… minding my own business, letting my mind just float through the day's events, enjoying the peace and quiet twilight brings… when I spied out of the corner of my eye a fat, gray and white, fuzzy field mouse sitting on the porch rail beside me. I looked at him… he peered at me….

"Are you freakin kidding me? Where are you all coming from? Is there a Mouseville somewhere close by? Go HOME!" Yes, I have now resorted to having conversations with these mini hordes that are determined to take over my house… and no, I did not expect Mr. Mouse to answer me back… what I expected was for him to run… run… run far away… but he didn't.

He just peered at me, blinking his little eyes… twitching his little tail… wrinkling his little mousy nose….

Disgusted I got up and made my way back into the house… only to see, to my astonishment, this freaking little bandit following me… which of course led to the dance... of me jumping around hoping he would not skitter up my legs… which also sent the dogs into a frenzy... which also sent hubby into a tirade of hollering from the bedroom… which also sent me into a fit of laughter at how crazy this household has become and how brave these little terrorist mice are….

Dogs jumping, me jumping and swatting at the mouse, throwing slippers and boots and basically anything within reach at the mouse… chairs knocked over, hassock slung across the room, one brave, lone mouse causing more havoc and chaos than a two-year-old child on a sugar high….

I give up… I have decided that now I will give the house over to the animals and move… far… far… far away….

And, I am leaving hubby, the dogs, the cat and the mouse behind…. I will take my Little Chicken Mafia, The Little Nasties and of course, Peckerhead….

But seriously, I am plotting and planning on how to get rid of the pesky little pellet pooping mice without having to resort to all-out war tactics.

Meltdown

Good morning. It's been awhile since I've shared a story from the homestead. Work has had me busier than a one-legged man in an arse kicking contest. And this winter's last storm has left me and the prisoners scrambling and stressed out… the Little Chicken Mafia refuse to leave their cell, the Little Nasties have learned that swimming in the deep snow is almost as much fun as swimming in their pool and Peckerhead, well he just plain out is one miserable little rooster being cooped up in his cell with five females to drive him crazy.

Yesterday's snowstorm on the homestead was the icing on the cake of a very long, tedious, cold, bitter and hard winter. Amidst the snow and the fact that our plow truck ended up in a ditch and topping that off with my only snow shovel breaking… I was ready to throw in the towel and thoughts of moving south ghosted the perimeter of my mind. Tromping through over a foot of snow to carry heavy water pails to the prisoners, slipping and sliding as I hustled arm loads of wood over a snowbank and up onto the porch, hubby swearing and yelling as he and my son worked on getting the truck unstuck was almost too much for me….

I did what any sane and normal person would do…. I had a temper tantrum that beat all temper tantrums…. I threw the broken shovel across the door yard, I yelled at hubby to suck it up; I kicked the water pail across the yard and I threw a hunk of wood at my front door breaking the screen and scaring the dogs half to death. I stomped my feet, bitched to anyone who would listen and told hubby we were selling out….

Then I made my way into the house, stoked up the fire, made a cup of hot cocoa and quit life for about two hours…. I didn't care that the truck was stuck…. I didn't care that the snow continued to fall, I didn't care that my water pail was somewhere in the back yard getting covered with snow and I didn't care that I couldn't shovel the front steps…. I had cocoa; I had a warm fire and you know what? It was all good.

This morning we have over a foot of snow that I will have to wade through, we have a beautiful morning sky with the sun just rising and I can hear Peckerhead singing in the coop… so quitting life for that two hours was okay for me… okay for the homestead... the truck is out of the ditch; the animals are warm and happy… and me? I am preparing for the next storm… and buying a new shovel.

Snow Fatigue

I am snow fatigued. Yes, there is a condition in the North East that is snow fatigue and after three Nor' Easters in the span of two weeks I've had enough. On this morning hubby is out, bundled up in a snow suit, pushing the blasted white stuff out of the driveway… I am still in my pajamas and wrapped in my blue and white Eeyore bathrobe. I can hear the prisoners squawking for breakfast and I dread the thought of slumping my way through two feet of snow, carrying a water pail, to feed and water the little buggers….

So, tired and grumpy I yell to hubby and in my sweetest voice, convince him to give me the morning off and take care of the squawking little beasts. Peckerhead has been crowing since before daylight, the Little Nasties quacking loudly as his backup band. And it continues to snow. And I am finding myself angry about it all. I want spring… the winter has been too long by months already…. I dream of warm misty mornings, coffee on the front deck and watching the sun rise. It is a nice dream.

Hubby slogs his way through the snow after I give him careful directions on what I do to settle the prisoners. Fighting through the outer pen gate, which is icy and frozen, he fills the Little Nasties water dish. He can't close the outer pen door… it is frozen to the ground. I watch from the front porch with a grin on my face… he doesn't do prisoner duties often and now he is getting a taste of what I've been dealing with all winter.

Next, he makes his way to the coop where I hear a ruckus going on. I wonder if he's walking into a riot… the prisoners sound awfully rowdy in there. I hear him talking and listen in closely.

"Get back!"

Squawk!

"Move you idiots so I can get to the dish!"

Squawk!

"What the fudge! Get off of me!"

Squawk, squawk!!!!

"I'm going to punt you like a football if you come at me dickhead!"

(Here I am guessing with a smile on my face that Peckerhead is ruffling his feathers and threatening my husband. The thought makes me chuckle.)

I suddenly hear a crash and then hubby lets loose with a string of threats, swears and grunts.

"You little Bast****! I'll show you!"

More crashing greets my ears and now I am just about rolling with laughter as I watch hubby quickly let himself out the coop door looking a bit hassled and hurried.

He looks at me and shakes his head.

"When was the last time you fed those little beasts? They attacked me getting to the food for God's sake!"

Laughing, I followed him into the house.

"Every day, twice a day. They are not starving although every morning you would think they are by the way they attack their food dishes."

He shook his head.

"You need a baseball bat… that fudging rooster is mean! I thought he was going to take my dang head off and the others would be picking my bones bloody!"

Smiling I nodded…. Yes, they are unruly little creatures… unpredictable and often plotting and planning their human caretaker's demise in probably the most gruesome of ways.

"Ummm, did you happen to gather the eggs?" I asked with a grin.

"Fudge you!" He replied.

So, needless to say I did have to venture

through the mountains of snow… my breakfast was not going to gather itself….

"Bring a baseball bat." Hubby warned me as I slid on my pink muck boots and headed out the door.

"Awwww… honey, I got this covered." I said as I grabbed the broom on my way out.

There is no taming of the beasts in Broadley Penitentiary… only quiet and uneasy chaos… but I am still top dog….

Normalcy Is Over-Rated

I truly believe my animals have identity issues. Especially my cat, Kitty Girl, who I don't think believes she's a cat, but rather, a princess. Now most cats will normally hunt and chase anything smaller than them, chipmunks, squirrels, mice, bugs, you name it… I once even had a cat, an old Tom, who'd bring me snakes and lay them on my doorstep… but not this Kitty Girl… she feels that it would be beneath her to hunt and chase….

Lying in bed last night as she curled up on my chest and purred in my ear, she and I had a talk. I could hear the mice chewing in the walls and skittering back and forth, and Miss Princess just sat contentedly, ignoring it all.

"You are not a normal cat!" I hissed.

Purr ... rumble… purr….

"What the fudge is the matter with you? You know, the dogs are putting you to shame girlfriend."

She looked into my eyes with her beautiful green ones and I swear she shrugged her shoulders.

"Can't you hear them in the walls…? I bet they're running around on the floor right now… taunting and teasing you…."

Purr… rumble… purr….

"You are such a lazy miss… get up and get those mice!" I hissed tiredly.

The gnawing was beginning to drive me crazy, and I threw her off the bed and went in search of the flashlight… Tara-belle… was prowling… she could hear the mice too… and they are brazen little poop machines….

"You are not a cat! You do not have a prayer of catching them!' I growled at her…. My dog (see identity crisis) thinks she's a cat. My cat (see identity crisis) thinks she's a princess. Tara-belle, although she tries hard to catch the running, frenzied little fur balls…. hasn't got a chance… in fact, she's more like a walrus in a kiddie pool, flopping and clumsy which ends up disrupting furniture, breaking things and amazing floor slides as she runs like hell all over the house.

I open the cabinet door and sure enough, Mr. Mouse is staring back at me…. I whack at him like the game Whack a Gopher, and he skitters away…. I chase him to the drawer underneath the cupboard and there he is, nestled among my dish towels and peeking his beady little eyes out at me… terrified… I take a swing at him again with my orange flashlight… finding I am as clumsy as Tara-belle when it comes to catching mice….

Kitty Girl saunters out to see what I am doing, and I round on her angrily….

"Will you start acting like a fudging cat and

catch these freakin mice?"

Sitting back on her haunches and wrapping her tail around her, she lifts a paw and licks at it as if she's bored. I get the sense she's telling me to Fudge off… in her snotty, feline way.

I threaten her… I yell at her... and I just get the look.

"I swear, you are going to be a barn kitty… I mean it! I will move your arse out to the barn and the only food you will get is what you can hunt and catch!" She gazes at me and yawns….

So, throughout the night I am chasing mice… playing Whack a Mole… because I know as sure as there is snow on the ground, that if I go to bed, I will be woken by a mouse crawling across my face, or in my hair, or God forbid, under the covers….

I need a cat who doesn't have an identity crisis… who can solidly and confidently identify as a cat and do what cats do… hunt and kill mice!!!!!

Penitentiary Broadley

I officially smell like a goat!!!! And I love it!!!! Set out early this morning to get the barn ready for the introduction of Willow and Daisy, my two new mini goats… moving pallets that were frozen in 4 feet of snow, laying deep hay for bedding, tacking a laundry basket to the wall and filling it with hay for them to eat…. The Little Chicken Mafia, The Ugly Nasties and Peckerhead all watched with curiosity… as they were counseled earlier on the newest arrivals. For weeks I stressed about how to transport these two fat girls… and yesterday my sister came up with a solution… we would use my nieces Suburban and put them in the back… well, sounded easy right?

Not so…. With my nieces husband driving, my sister in the backseat, and me in the way back we loaded in the two new prisoners… who proceeded to sit on my lap… while both simultaneously are trying to crawl up under my shirt for attention and comfort…. what a zoo crew we looked like as we drove the half hour drive wrangling two 60-70 pound, nervous, scared goats…. Laughing all the way… my great-niece, Jelly began, trying her hardest to feed them orange juice from her car seat, my sister hanging onto one of the girls for all she was worth to keep me from getting crushed, Bobby, my nieces husband yelling at every turn….

"Turn, hang on! Bump! Hang on!"

Oh, the joy of it all. Getting the two, Willow and Daisy, back to the homestead, we introduced them to their new cell mates… and Peckerhead was not impressed… squawking and growling he wrangled up the Little Chicken Mafia and flew the coop… The Little Nasties quacking angrily ran out the door and into the snow with me hot on their tails… and the two girls, bucking and pulling, ran right for the outer pen, crying and balling as they inspected their new digs….

My sister was sitting in the snow laughing so hard at the chaos and frenzy I thought she would have a stroke while I am chasing all manner of feathered fowl… and four-legged creatures while trying to keep everyone from escaping….

And finally, I have got everyone settled… for the moment… for this moment… until feeding time...

and by then, hopefully, everyone will have established their places in Broadley Penitentiary….

So, this day has started with squawks, crows, growls, bleats and quacks… and lots of laughter….

Prison Riot

The past two days here at Broadley Penitentiary have been interesting to say the least and with the onset of yet another snow event, I am feeling the extra work with having the Frolicking Frisky Fat girls amongst the prisoners. The Little Nasties have gone on strike and not laid their little ducky eggs for two days, the Little Chicken Mafia have been practicing their own brand of payback and Peckerhead… well he's just being a peckerhead. Everyone in the cells are protesting the crowded conditions….

The Frolicking Frisky Fat girls are loving, fun, and so curious… nosing the other prisoners creating a flurry of quacks and squawks… outrage from the fluffy feathered fowl… they charge the door each time I enter, all sixty to seventy pounds of them trying to make a quick prison break… and I've been bribing them with Ritz crackers so that I can inspect their fat little bodies and become familiar with their goaty ways. This makes the others jealous… and they vocalize angrily when my pockets are empty of extra treats….

So, as payback the Little Chicken Mafia has decided that they will roost above the two fat girls and proceed to drop poop bombs on them throughout the night… leaving me in the morning to have to wrangle the fat little butterballs and give them each a sponge bath. One of the chickens, Gladys, has decided that the best way to get me back for introducing this chaos into her life is to start pecking at the eggs the other cell mates have laid… leaving me two mornings in a row without my breakfast.

And Peckerhead, well he is just downright piffed off! Twice on my forays into his territory this morning he squawked and growled, puffed up his feathers and made a run at my legs… which landed him in solitary for the rest of the day… no bread and water for that little bad boy.

As with any prison, introducing new cell mates leaves a bit of chaos in its wake… normally there are fights for positioning… who is the biggest, the baddest and the orneriest prisoner on the block. I've removed anything that can be remotely used as a prison weapon, shivs, sharp sticks, nails, and paperclips…. And my solitary confinement cell is ready and waiting for the little line of guests that may be assigned there for a temporary stay… the first of course which is Peckerhead. I am hoping the Little Nasties will get used to the new guests and start laying eggs again for me and Gladys will stop trying to destroy the other hens eggs… and most of all, I hope the Little Chicken Mafia stops dropping poop bombs on top of the Frolicking, Frisky Fat girls. And Peckerhead… well… I guess that Bad Boy of the Barnyard now has some bigger and badder competition for king of the prison cell block… so we will see… tee hee….

No Sleep on the Homestead

I sat on the front porch this morning and watched the sky lighten. The sunrise hues of pinks and blues indicated the beauty of the day ahead. After the rough night I had, it was a welcomed sight. Yesterday, for the first time in many long weeks, I had made it off of the mountain and out for a shopping date with my sister. I was excited to buy a new vacuum and, yes, it takes very little to make me happy and the purchase of a new vacuum made me happy. After we shopped, we did what sisters do and stopped for a nice lunch… and this is where the trouble began. You see, my tummy does not do well with fast food and greasy fast food is the kiss of misery for me. But regardless I gave into temptation and had a deep-fried fish sandwich… and no, I couldn't stop there, I had to add greasy fries to that order, really asking for trouble.

So this time it wasn't the Barnyard critters that woke me up, it wasn't the mischievous mice nor the clumsy flailing's of Terra-belle… it was a belly ache that kept me tossing and turning well into the wee hours of morning… a belly ache to beat all belly aches….

I lay in bed moaning and writhing…. Misery wracking my body….

"What's wrong?" Hubby growled.

"I have a belly ache."

"Go poop!" he muttered. I could sense his frustration. He had to get up in a few hours and my tossing and turning was keeping him awake.

"I don't need to poop… go back to sleep." I moaned.

"I would love to go back to sleep, but it's kinda hard with you moaning and tossing and turning like a dying fish." He replied.

"A dying fish? Where the fudge do you come up with these things?" I hissed. Throwing the covers off, I stumbled out to the kitchen for some Maalox and an ice pack for my tummy. Opening the freezer (of course I didn't turn the lights on) I fumbled through frozen foods trying to feel my way to the gel ice pack I kept in there just for times like this. Dislodging all manner of frozen meat, I felt the pile start to let loose… and out came a frozen chicken… and of course… it landed on my bare foot….

"Fudge!" I screeched as I hopped around in the dark kitchen holding my foot, doing the one-legged dance with tears of pain trickling down my face….

"What is going on?" hubby yells from the bedroom… if he had fallen back to sleep, whelp… he wasn't sleeping any more….!

"Shut up already! I dropped a frozen chicken on my foot… I'm okay!" I yelled back… limping, I picked up the frozen bird and shoved it angrily back into the freezer which caused a cascade of frozen Hamburg, frozen peas, frozen chicken fillets to come tumbling out… everything but the fudging ice pack I was looking for….

"Oh, MY GOD!" hubby yelled….

"There is no God!" I hissed angrily… ... (Of course, there is a God… but hubby just peeved me off and I could think of nothing else to screech back at that point).

"Will you just come back to bed before you wreck the kitchen and break a bone?" he pleaded….

"No! I'm sleeping on the couch!" I yelled….

All of this at 2 a.m. and now I am sitting on the front porch, drinking my coffee, ice pack on my foot and greeting the dawn… Ohhhh my life….

Are You Kidding Me?

Daisy, my new Nigerian Dwarf, my little fat girl, started crying out at the butt crack of dawn. So, before I even had my first sip of coffee, I slid on my muck boots, still dressed in my Eeyore Robe, I set out to the barn to see what was bothering her....

Grabbing my little lantern, I made my way through the dark… her cries getting louder the closer I got. Entering the barn, I could find nothing that was out of the ordinary… Peckerhead and the Little Chicken Mafia were perched on their roost, the Little Nasties cuddled together on their nest sleepily. And Willow, Daisy's sister, was curled up on her bed of hay in the corner. Daisy stood there and bat her baby browns at me… rubbing up against my legs, rumbling in her throat.

Giving her a pet and telling her to quiet down, I went back into the house to fix my coffee. The minute I closed the barn door behind me I heard her start up again… a constant crying and bleating….

Hubby: "What is wrong with her?"

Me: "Danged if I know… I gave her food, water and fresh hay… she should quiet down."

Hubby: "I hope so… thank God our neighbors are a ways away... they'd be calling the cops on us about now… it's freaking 4:30 in the morning!"

Me: "I know what time it is… and your point?"

Sitting out on the porch, away from my grumpy hubby, I listened to her nonstop bawling… puzzled as to what was going on. Carrying my coffee, I made my way back to the barn, tripping and slipping through melting snow and mud that was as greasy as all get up and hoping I wouldn't land on my arse as I made my way through the dark. For an hour I sat in the barn with that little fat girl, petting and soothing her, urging her to quiet down. The morning light was just breaking the horizon when I made my way back to the house to get dressed for the day.

"You're spoiling that brat."

Me: "So."

Hubby: "That's why she's making so much freaking racket… you're spoiling her!"

Me: "Okay, so what do I do jackass? Let her sit out there and howl all morning long?"

Hubby: "Ignore her. She's just being a brat."

Me: "How bout I just ignore you, asshat!"

The crying was lasting for hours… no matter what I did, pet her, brush her, sit with her… she just cried and paced back and forth, in the barn, in the pen, on the leash (yes, I did take her for a walk on the leash)… finally I got on Facebook to my goat group and sent out my own cry for help.

Me: "Help, my goat won't stop crying and I don't know what to do!"

Reply 1: "What do you mean crying? Explain exactly what she's doing."

Me: "She's been crying since early this morning non-stop and she's driving me crazy… I have fed her, watered her, brushed her, pet her, and took her for a walk… I don't know what else to do."

Reply 2: "Does she have a temp? Is her nose runny? How do her eyes look? How's her poop?"

Me: "No runny nose, eyes look fine, pooping pellets like normal and how do I know if she has a temp? Do I feel her nose? Like a dog will it get dry and hot?"

Reply 2: "Ummmm… no… you have to take her temp."

Me: "How?"

Reply 2: "Rectally."

Me: "Ummmmm what?"

Reply 2: "Yes, rectally. Lube up the thermometer and shove it in her arse…."

Me: "Are you serious? You aren't serious, right? April fool's joke? Please tell me you are joking."

Reply 2: Ummmm… nope, no joke.

Me: "Is there any other way? Really, I do not want to play with my goat's butt…."

Reply 2: "Suck it up… take the temp. I'll wait for you."

Soooooo… back out to the pen. Thermometer in hand, lube (bag balm) was all I had. Wrangling up to the fat white screaming girl I pinned her against the wall… her 65 pounds pushing against my 120 pounds… I wrestled, I pushed, I used my knee to keep her pinned and I finally got her temp… and God I felt like the world's biggest pervert!!!! Going back into the house, I jumped on the computer.

Me: "Her temp seems fine."

Reply 2: "Okay, so how old is she?"

Me: "Four."

Reply 2: "Has she come into heat yet?"

Me: "I don't know… how would I tell?"

Reply: "Well, she'll get real lovey with you, pace

a lot, swish her tail frequently, and there will be a clear, stringy, snotty discharge coming from under her tail."

Me: "Oh, nooooo… I ain't going there! Nope, nada, no way!"

Reply: "Oh, for Pete's sake… just go see if you can see any discharge…."

At this point I'm wanting to puke… seriously… my coffee was threatening to come up…. trudging back out to the barn I approached Daisy once again… now the Chicken Mafia were all awake and clucking for their breakfast, Peckerhead was eyeing me menacingly and the Little Nasties were quacking in protest for still being locked in their cell… and Daisy, well she was running and skirting around me not wanting a replay of the thermometer and butt action….

Cornering her, I pinned her again against the barn wall and lifted her tail to inspect the nether region…. Willow, her sister, was nudging my butt with her head and the Little Chicken Mafia were crowding around my legs and pecking at my boots… yelling, I scooted the little carnivores away from me, pushed Willow who decided to head butt me back and went in for the look….

Ewwwww….

Back in the house I jumped back onto the internet.

Me: "Yes, there is some gross stuff dripping from her."

Reply 2: "Bwahahahahahha… she's in heat… this will last about three days… no worries… she'll be fine."

Me: "I am just going to shoot myself…."

Reply 2: "LOL… just wear ear plugs."

Me: "Thanks… thanks a lot."

So, I have a goat in heat who I find out will be having stringy ooze dripping from her butt and screaming for the next three days…. Oh, heavens… my life just got a lot more interesting for sure.

Ain't No Project like a Homestead Project

After having my morning coffee on the front porch watching the sky turn from inky blue/black to a pale overcast, I decided that today would be perfect for some outside work that had been on my list of things to do all winter. The temps were going to be cool, but the weatherman promised sunshine. So, getting my morning chores done quickly, I set off for town. I needed to get food for the ever-hungry prisoners and wanted also to get more fencing to extend their roaming area.

Thus, late morning found me unloading a fifty-pound bag of dog food, a fifty-pound bag of chicken laying pellets, a fifty-pound bag of ducky food and 2 bales of hay along with a fifty-foot roll of fencing. This fencing was determined to kick my skinny little arse right from the get-go… it was nearly as tall as I am and God, I swear, just as heavy….

Now, the prisoners were feeling a tad bit rambunctious this morning… Peckerhead was in the outside pen away from his girls, the Little Chicken Mafia were free roaming the yard and the Little Nasties were inside the pen with the two Frolicking Fat girls along with a pissed off rooster…. What could go wrong?

Add me to the equation, trying to run fencing, fend off the beasts and it was of course a recipe for disaster. As I'm in the pen hauling the fencing, hammer in my hand, nails in my pocket, Peckerhead decides it's time to show his wrath…. He charges straight for me, feathers puffed up, neck stretched out, growling and a deadly glare in his eyes… I did what any sane person would do when faced with an angry rooster… I threw the fencing at him… the whole roll….

"Not today you little Flucker!" I yelled. This backed him off momentarily… picking up the fencing I began once again to unroll it and run it along the tree line… struggling every step of the way… an angry rooster glaring at me… two goats following me like puppies… the Little Nasties quacking loudly demanding more food… and my dog chasing squirrels in the woods….

Pinching my fingers numerous times in the fencing… Daisy decides to use that moment to bawl out her horny frustration, right in my ear… and then proceeds to jump on my back… a sixty plus pound horny goat…. trying to hump me as I am back to, bent down and driving a u-nail into the fencing attaching it to a tree…. I am laughing so hard that I'm crying as I am being molested by my goat…. Meanwhile, her sister, Willow, comes over and decides it's a good time to nibble on my ear… and Peckerhead is circling us all, growling and posturing, knowing full well that he has the advantage… if he attacks now, I am screwed seven ways to Sunday….

Pushing the horny goat off of my back, slapping at Willow to let go of my ear, I turn and face

Peckerhead….

"I do not think so!"

Reaching out a gloved hand I grabbed that little blasted bad boy by his neck and tucked him under my arm like a football….

"We are gonna get something straight right now!" I yelled as I held him tight….

"I am going to lay this fencing without getting attacked by you, molested by a freaking goat and nibbled on by her sister… you hear me Peckerhead?" With that I tossed him back into the jail cell and confined him for the duration. Turning to Daisy who was eyeing me again with a lusty look, I grabbed her collar and drug her into the barn… and that is where she stayed, bawling out her unhappiness until I finished my work.

Finishing the fencing I sat on the front porch and waited for the lunch time call from my hubby… I was feeling pretty satisfied with the work I'd gotten done and the sunshine felt good on my face.

"Hey, what'cha doing?" Hubby asked.

"Just finished getting molested by Daisy and attacked by Peckerhead." I replied.

"And you?"

I hear a chuckle on the other end of the phone.

"Okay, what do you mean molested?"

Me: "Daisy… she's moved from bawling to humping… I made the mistake of turning my back on the horny little beast."

Him: "You're insane… what did you do?"

Me: "Tossed her ass off of me and confined her and Peckerhead to their prison cells… got my fencing finished though."

Him: "Why do you pick on Peckerhead?" (My

husband loves this nasty tempered little rooster and always thinks I'm pecking at him).

Me: "What? Picking on him hell… the fluffy little flucker is always challenging me… he's lucky I don't put him in a stew…" I sputtered.

Him: "Did he draw blood?"

Me: "Nope… didn't give him a chance!"

Him: "Then I guess you've had a good morning then…."

Me: "Well thank you for your concern… asshat!!!!!"

So here I sit writing this story for you all, I've got my fencing finished, my critters are happy, no blood was shed, and no animals were harmed in the making of this story…. And now I am going to bake me some brownies because this lady needs and deserves some chocolate.

The Struggle is real

Today was a sister day… the very best kind of day up here on the homestead. This morning I had eggs to deliver and clothes (hand-me-downs) to pick up at my sister's house. So, I set off early after feeding and watering the hungry little horde….

Sitting and visiting, having coffee I asked my sister what she was up to for the day and seeing that she had no plans, I invited her up to the farm to keep me company while I canned some beef I had picked up on sale at the store. I love the prepper side of homesteading and it always makes me feel good when I can and put food up for those days that I either do not have the time nor the energy to cook from scratch. Now, my sister knows that here on the homestead one never sits still. There is always something needing doing… so after I put the roast in the oven to brown it, we headed out to the barnyard to tend to the animals. I had been wanting to trim the Frolicking Fatso's hooves but couldn't do it alone… so now I had the extra set of hands I needed to get the job done. So together we wrangled the first little fatso, Daisy, and while Chris (my sister) straddled her I lifted her leg and began to cut away the excess hoof….

The laughter and giggling erupted… the bucking, twisting and kicking began… and this 60-pound fat girl was knocking us two grown women all over the place.

Me: "Hold her!"

Chris: "I am!"

Me: "Hold her tighter!"

Chris: (Profanity ensues here… we'll clean it up though.) "What the fudge! Hold still, Daisy! God blessed! Do we need a break?" (This is only after five minutes….)

Me: "Nope, I almost got it!"

Daisy: Bleating and struggling, thrashing us around.

Chris: "Son of a b*****!" (and she hoists the ass end of my goat up in the air).

Me: "Okay… just hold her now!"

Well, it took us probably twenty minutes to do the little Fat girls' front hooves… twenty minutes of cursing and struggling which left us spent and winded.

Me: "Okay, let's take a break and have a cup of coffee…"

Chris: "Tea."

Me: "What the fudge ever!"

So, in the kitchen we sit, having our drinks and her sipping down teaspoons of my Elderberry syrup. I introduced her to my latest herbal creation… Transition Time Oxymel… she is always game for a taste, so I pulled out a big soup spoon and filled it up.

Chris: "Is this like the fire cider?"

Me: "Yeah, a bit, but not as strong I think… just down it in a quick gulp." (That should have been a dead giveaway-her first clue).

She takes the spoon full I gave her and starts choking and sputtering… eyes filling with tears and her face turning red as she's trying to catch her breath.

Me: "Oh, I forgot to tell you, it's got quite a bite…. Bwahahahahahahahaha"

I thought I was going to pee myself; I was laughing so hard… and she was ready to throttle me… as I choked, sputtered, snorted and laughed.

Me: "But it is sooooo good for you… teehee…."

Her: Again, with the profanity….

Once we finished our tea, we headed back outside to tackle the other little Frolicking Fat girl… Willow... and thankfully that went smoothly. Chris helped me feed and water all the little critters and give the two goats some exercise by walking them around the backyard on leashes. Meanwhile, forgetting to close the fence up, one of the Little Chicken Mafia prisoners made a jail break… so chasing a chicken was the next on our agenda… and Chris volunteered to give the chase… now I know my prisoners and those little cluckers will give you a run for your money….

Off through the snow and woods this fluffy

little clucker ran... squawking and flapping all the way… and my sister hot on her tail, yelling and bitching the entire time… and of course I am howling with laughter at the site of them….

"C'mon sis ... ya gotta be faster than that…" I yell across the yard….

Chris: Again, with the profanity… she has such a trashy mouth….

Me: Laughing and standing at the fence….

Her: Running after my chicken….

Chicken: Running and clucking… taunting her….

Me: "Bwahahahahahahahaha!"

Well, she finally caught that fluffy little flucker and threw her over the fence back into the pen… and she's breathing so heavy that I think my chicken has killed her….

Back into the house, cleaning up the canning supplies I am preparing to bring her home. She's had a great time on my little homestead, and we've had a wonderful visit… but the fun ain't done yet…. Ohhhh nooo… never done at this little asylum….

"I have to lock the prisoners in before we leave."

Her: "Okay, I'll help."

So out to the pen we go… and as I am piling stuff into the car, she opened the fence to collect up the water pans and feed dishes… somehow she got snagged on the fence… Willow, my fat little goatie girl barreled into her, knocking her away and escapes out of the fence… Daisy, her sister, follows quickly behind… and Chris yells….

"Nancy the goats are loose! Fudge!"

I stand in the front yard playing wide receiver

trying to wrangle Willow while Chris has Daisy by the collar and trying to untangle herself from the fence… I am of course laughing… she's bitching ... the goats are determined to drag us both to kingdom come and it starts at that moment, snowing… my sister looks at me, I look at her and we both look up at the sky….

"It's fudging snowing!" Chris yells.

Me: "Yup, only up here on Twisted Hill, we're in God's country now!"

Chris: "Screw this! Take me back down off of this mountain! I hate the snow!"

On the ride home we both were laughing at the antics of this farm life… completely exhausted but happy, full of fresh air, farm animals and good company….

Yes, this is my life on the homestead… and now my sister knows… my stories are real life… not Memorex….

Sentencing

Well, there is a somber mood on the homestead tonight. As a homesteader there are times when you have to make the hard decisions, and tonight I made a hard one…. Peckerhead will be no more tomorrow morning. This afternoon I heard a squawking and screeching in the exercise pen… and found one of my hens of the Little Chicken Mafia, beaten and bloody... with Peckerhead standing atop of her. There was a rumble, and he had gone too far….

Now as you know from my stories, Peckerhead and I have had a challenging relationship… him strutting and threatening... and me, never able to really trust turning my back on this little troublemaker. He's got the other prisoners in a frenzy time and time again… enticing them to jail break… enticing them to fight amongst each other… and twice now he's drawn blood in the midst of his battles with the female prisoners… and twice I've commuted his death sentence… but this time, he took it too far. Not only did he severely hurt one of the girls… he drew blood on me, the warden… and this I won't tolerate.

So, tomorrow morning, bright and early, I will capture him one last time. I will send up a prayer of thanks for the life he lived and for the food he will put on my table….

R.I.P. my nemesis and challenging little convict. Your fluffy little flucking arse has had a good run… if there is a heaven for fluffy little fluckers like you I hope you have an abundance of bugs and wild foraging, I hope you have an abundance of sassy little chicks to gather round in your flock.

Jail Break

It was 4 a.m. when I heard the first crows of Peckerhead, my flucky, feisty Bad Boy of the Barnyard making his presence known. Groggily I climbed out of bed, tired and grumpy. Should I wear black today? The colors of the Executioner? Black would fit my mood. Sitting on the front porch, I drank my coffee and built up the nerve to do the dastardly deed.

Hubby: "Your arm is a mess. You better wear gloves when you grab that little jerk."

Me: "Yeah. No worries, if you come home tonight and find me face down in the pen laying in chicken poop and the birds pecking at my bloody body, you'll know I lost the battle."

Him: "Drama much honey?"

Me: "Fudge you... go to work." You see, I am a little grumpy this early in the morning. After the coffee starts working, I become a nice human again.

Wanting to get the dastardly deed done early, I decided that his execution would take place at the first rise of the sun... even before the other prisoners got fed and watered. I could hear the Little Nasties (my duckies) inside their cell quacking and demanding breakfast, I could hear the Little Chicken Mafia squawking as they pooped out my breakfast and the Frolicking Fat Girls (my goats) were snorting and stomping in their stall waiting for their morning sweet feed and grain. The prisoners are early risers and they have this warden trained to come to their beck and call. But not today my prisoners, today the Warden had to carry out a rather distasteful deed.

Throwing on a ratty pair of jeans, my pink muck boots, my hubby's welding gloves and yes, his motorcycle helmet, I took the long walk to death row where Peckerhead awaited his fate. I'm sure I looked like a confused two-year-old with an identity crisis, but I didn't want to be bloodied up again by his dangerous beak and claws… normally I feed and water the prisoners in my Eeyore bathrobe and pink muck boots….

Opening the cell, I cautiously walked in.

Peckerhead glared at me.

"Sorry little dude, you brought this on yourself."

A small growl erupted from his throat and his neck feathers fluffed up… he was on the defensive... and getting ready to attack.

"I don't like this anymore than you do so suck it up and be a freaking man!" I hissed as I made a lunge toward him. He squawked and screeched, and the battle ensued… I dodged as he made a run at my legs… and I felt his beak hit my boot… I kicked him off of me and dove in like a linebacker for the winning tackle… and that little flucker jumped up and over my shoulder, hissing and flapping all the way. Spinning around I lunged at his tail feathers trying to get my hands on this little pterodactyl before he made a run for it….

Now seriously, have you ever tried to catch a rooster hell bent on escaping? Have you ever tried to catch a rooster with welding gloves and a motorcycle helmet on? It doesn't freakin work!!!!!

I chased him across the yard, him squawking, me yelling and looking like I belong in some padded cell somewhere… helmet off and tossed across the yard, Peckerhead running… me winded and determined to catch that flucking fluff arse…. I can hear the Little Chicken Mafia laughing… and pointing at me as I make a spectacle of myself… I can hear the Little Nasties rooting the dastardly little bad boy on….

It is too fudging early for this either that, or I am just getting to dang old….

So today, I will be plotting and planning on how to catch this runaway rooster… and watching my back as I work in the yard… because I know it's either going to be him or me… and it's do or die….

The Verdict

The fugitive, aka Peckerhead, has been apprehended. He is now in Solitary Confinement and his rights to General Population have been revoked. At 0200, a jury of his peers met, Marianne and Ginger, Betty, Laverne and Shirley…. Marianne, with the help of her cell mates, hobbled to the stand and gave a strong and compelling testimony as to why the accused deserved the Death Penalty. The jury met behind closed doors for several hours before presenting to the Judge, the Honorable Nancy, with a split decision. Ginger had to leave to lay an egg, so the Jury was excused until 0900 tomorrow morning. The prosecution gave a strong and compelling case against the accused, aka Peckerhead. The defense rifled back with a strong and compelling evidence of their own. Until tomorrow, Peckerhead will be held in solitary with the option to have a visit from the local clergy.

Court Adjourned.

Spinning World

This morning was a rough start here on the homestead. I woke up with vertigo… now for anyone who suffers from this you know the drill… your feet hit the floor, then you do….

So, after picking myself rather ungraciously, I stagger out of the bedroom holding on to furniture, walls, my dog, hell I'd of even held onto my worst enemy at that point. Hubby looks at me and shakes his head….

"Vertigo?"

"Well, duh… no I just drank too much already this morning."

He fixed me a cup of coffee and helped me to my chair out on the porch where I sat in a constant spin for a half an hour.

"Are you going to be okay?"

"Mmmmhmmm…."

Staggering around the house like a drunken pole dancer not worth more than a dime, I managed to brush my teeth and throw on some clothes… no minor feat considering the condition I was in….

I was late getting to the prisoners and the chorus calls from their cells were letting me know loud and clear they were not happy with the warden this morning. So first to the chicken coop where the Little Chicken Mafia were squawking and threatening to riot… slopping my pail of water all down my legs, spilling pellets of grain in a trail behind me like Hansel and Gretel in hopes I'd find my way back to the house without landing face down in the dirt.

Then to the Prisoner, aka Peckerhead, who glared at me as I entered his cell. He was not a happy prisoner this morning, and I knew to tread lightly… as nauseous as my stomach was from the vertigo, I knew I just wouldn't be able to take the blood if either of us lost our temper…. Surprisingly he let me into the cell without incident.

Then off to the Frolicking Fat Girls and the Little Nasties… and of course, the minute I entered their cell, I stepped in a big gob of duck poop… me and duck poop and vertigo… equal no balance… as I landed on my arse… grain flying everywhere and the water bucket emptying on top of me…. Ohhhhh does it get any better than this…????

Well, yes, yes it does.

Thinking I was in the home stretch and I could crawl onto the couch to rest and see if my head would stop spinning… I saw my Fat Little Girl, Daisy, escape her fence line… so staggering, running, loping and dizzy I took off after her… down through the back field, out through the woods, through brambles and branches… screaming….

"Dear God in Homestead Heaven do not let me die!!!!" because honestly folks, it wouldn't have taken but a breeze to knock me off my feet. Finally catching the Frolicking Fat Girl, I led her back to the prison yard. I poop you not. Five minutes later she was gone again….

We did this three times before I finally put her into solitary….

And drunken pole dancer danced her way out to fix the fence….

Lessons

These are the lessons learned from this homestead.

If there is poop to be found, you will find it. You will step in it, sit on it, lean your hands into it, have it splatted on you, slung at you, slipped into, fell into and wiped on you. There is no getting around the poop.

Never ever use power tools or any type of tool when you have vertigo… you will bleed, puncture, wound, stab, slice and drive something into some body part.

Always have plenty of band aids on hand. You will use them.

Never ever wear nice clothes onto the homestead. You will be bled upon, pooped upon, puked upon, spit upon. You will fall in the mud, fall in the poop, and fall into the pee….

You will encounter the unusual… bats, mice, possums, skunks, bees, wasps, and every other creepy crawling thing you can think of. Oh, and did I say bats?

Be prepared to be injured by your animals… you will be kicked, pecked, dive bombed, spurred, bitten, and chased. You will have bruises… and sport them like a champion. You will have a Peckerhead… you can count on it.

Never wear flip flops into the chicken coop or the barnyard… your toes will be mistaken for fat little grubs, they will be stepped on and pecked, and they will be stubbed and bloody.

You will start talking to God in the form of questions…. "Why God, Why Me? Are you serious God? Really?" or my favorite, "Dear God, please don't let me die!"

You will put your hands in the most unusual places, and this is usually some part of a body cavity of one of your critters….

You will become very used to the gross.

You will learn endurance… as you run, as you bend, as you twist your body into positions you had never thought possible.

You will talk about your day at the supper table as you hash out food and gross little tidbits from your day.

You will find yourself having barnyard breeding conversations with others… and think this is normal.

And lastly, you will thank God every night before you sleep that you made it through another day.

Pardoned

Today I feel I need to catch everyone up to date on the Peckerhead saga… and after experiencing two days of prisoner riots… which will be another story, I finally settled the jury down and talked to them about a verdict.

The verdict came back with a two–three split. The Little Nasties were excused from voting due to no interest whatsoever and the Frolicking Fat Girls just plain wanted to kick his feathery, flucking arse….

Judge Nancy reminded the Jury that whatever the sentencing, they had to keep it in accordance with the Geneva Convention as far as fair and humane treatment.

Two of the votes were for Death by the Broom Stick method, three votes for life in confinement.

So hereby are the results of Life of Confinement:

Peckerhead was found guilty on all counts of chicken assault/battery.

Peckerhead was found guilty on all counts of

Assault on the Warden.

Peckerhead was found guilty on all counts of orchestrating discord and riot amongst the other prisoners.

Sentencing:

Peckerhead must attend weekly Sensitivity Training… given by none other than the Warden.

Peckerhead must stay in Solitary for the next six months with no conjugal visits from any of the Hens.

Peckerhead is allowed two hours of exercise time once a day, time to be determined, in the outside pen alone.

Peckerhead must attend one Anger Management meeting weekly.

If Peckerhead fails to follow through with any of these stipulations set forth here today at the County District Court by the Honorable Judge Nancy and the Sitting Little Mafia, then the Broomstick method hearing will take place at above said Court.

Signed Honorable Nancy.

Court Adjourned.

Life

I write a lot about my days here on this little patch of homestead. I try to share the funnier side of this life. But today I have to tell you, it is so much more than the short, funny stories I tell. It is bits and pieces, parts and parcel of everything that pulls me out of bed at the butt crack of dawn each day.

It is the grace and gratitude of living close to the earth. Of living as simply as possible.

It is a lifestyle of sustainability. Of frugality.

It is about stepping into my kitchen and preparing a meal from scratch or preparing a medicine from the plants I love.

It is tending the animals who depend on me for their food, their water, their housing and their health.

It is fixing fences, building enclosures and repairing, constantly repairing.

It is tilling the soil and planting the gardens for our food.

It is about chasing chicken, wrangling roosters and body diving to tackle an escapee goat.

It is the cold and blustery winters where boredom can make you crazy and sad and breaking the ice on a watering trough for the umpteenth hundredth time in one day.

It is thc frigid wind and thc snow and thc icc that send you running back inside with frozen fingers and toes.

It is hauling your arse out of bed before the light of dawn because you know, the animals depend on you for everything.

It is fighting the wind and rain as you chase little nasty ducklings through the woods because they escaped their pen.

It is the beauty and wonder, the hardships and the laughter. It is the tears when an animal dies and it is the laughter when they greet you at the fence.

It is midnight visits to the barn to make sure everyone is tucked in and safe.

It is tending to wounds and illnesses… and praying that you are doing everything right.

It is looking at a task or a need and asking yourself if you have one more ounce of energy to get through it.

It is taking out the complications because you know, it is only as complicated as you choose to make it.

It is busted hands, pulled muscles, scratches and scrapes, band aids and blood.

But most of all… it is the wonder of every new day and the days that leave you wondering.

Mother's Day

As today is Mother's Day, this homesteader decided that she would sleep in a bit. So, the prisoners were a bit put off when their breakfast was late. So, put off in fact that there were chorus calls of rioting going on behind their closed cell doors. The Little Chicken Mafia were hollering and squabbling, the Little Nasties so loud that I swear my neighbors up the hill could hear them and Peckerhead, well, and he had been crowing and threatening me since about 4 a.m. As I walked through the prison hallways I heard taunts and threats, namely threats to call the Governor about their apparent mistreatment… but hey, it's Mother's Day and this warden deserved a little extra shut eye.

Feeling badly about their late breakfast, I decided to let the Little Chicken Mafia out of the pen and let them free range a bit. I was desperate to give them anything that would soothe their ruffled feathers. So, hubby and I sat at the picnic table and quietly enjoyed the morning with our coffee in hand and rifles on the ready in case one of them fluffy little fluckers tried to make a break for it. Coffee and fresh blueberry muffins were my breakfast of choice and as I watched the prisoners, I enjoyed a moment of peace in the early morning sun. I hear a snicker from hubby and looked down to find all of the prisoners gathering around my feet waiting for me to drop a crumb or two. Beady little eyes, cold and hostile, staring me down.

Hubby: "They are like a bunch of little puppies... just begging for a treat."

Me: "No, not fuzzy little puppies. More like winged raptors. Those little fluckers would pull a shiv in a heartbeat and slice my throat."

Hubby: "Aren't you projecting your fear onto those sweet little birds?"

Me: "Fudge no. It kinda goes like this my dear. Imagine these same little sweet prisoners at ten feet tall, sharp beaks and razor-sharp claws… they are carnivores. They eat meat remember? Now, they view you and me and our dogs and our goats and every other living creature as a food source. We would be the ones in the pen if nature had reversed itself and these sweet little birds would be our keepers. Now, as far as I am concerned, you'd be screwed and in freezer camp already, but me, me they would keep around a bit longer for breeding purposes. Because, I would produce them tiny little humans for food…but once I was past the breeding age? Then it would be off with my head and my bloody carcass would be in freezer camp too… now what do you think about those sweet little prisoners? Those fluffy little killers?"

Hubby: "I'm going in the house… you are so morbid."

I peered down at the ravenous little prisoners at my feet and smiled. Thank God I am bigger than they are… the thought of them being ten feet tall just sends shivers down my spine….

"Shoo you little carnivores…shoo…."

Broken Peace Treaty

Peckerhead…. Oh, my little dude, my feisty, flucking Peckerhead. I should have known; our peace treaty did not last long. But today, Ohhh, I was so ready for that fat little fluff butt. I entered the pen, solitary, to give the white beasty convict some food and fresh water… I should have known by the growls... the fight would be on. And yes, you can ask my dear friend, Gee about his growls… she thinks he's the Devil Incarnate… so anywhooooo… I ignored the growl….

But I was ready. I opened the solitary door… I took one tentative step then another… he glared at me and hissed….

Ready to rumble. Do the deed. Dance the dance. Pay the piper. He launched himself at me like a bullet….

I caught him mid-air and spun him off where he hit the wall and shook his head. Standing, feathers in full fluff, eyes trained on me….

He came back, hissing, swearing, and growling… claws at the ready, beak zeroed in on my face. I gave him a taste of my boot… he didn't like it.

Third times the charm and this time when he launched I caught him by his legs... which by the way wanted to kick my arse… and I hung him upside down… and he kicked... he yelled… he squawked… he beat his wings… he threatened me, my mother, my sister and my entire family line, but I held on as if my life depended on it... which it did.

When he finally heaved a defeated sigh, I flipped him up and hugged him… tight (maybe tighter than I should).

"You will love me!" I yelled as I walked him around the yard in a bear hug. For. An. Hour. Oh, the humility… I strutted him in front of the Little Chicken Mafia… I danced him in front of the Little Nasties… I pet him like a puppy… the Goatie Girls were laughing at him, adding to his humility. Even the birds in the trees above were chuckling on their perches… total humility for this Bad Boy of the Barnyard…. And. He. Hated. Every. Moment.

"You will love me." I said with each step. The look in his beady, cold eyes told me it would never happen. Not today Peckerhead, but eventually, you will love me!

Today's battle was won… but this bad boy I know will try again… and again… and again. But this girl will be ready, always on the ready.

Memories Made

Just had to share with you all because only you folks will understand how happy this makes me.... I had my 6 year old grand-daughter over yesterday and the weather was calling for a light frost last night so I had her help me pull in some of my more tender herbs growing in pots... well of course we had to pinch and sniff... right? So, the first plant was eucalyptus... and I asked her to repeat it back... well now she knows this plant as eucakisses... she pinched and sniffed... and says, "Grammy it smells like medicine... delicious!" Next on the agenda was my lavender, same thing, pinch and sniff... and Ohhhh she loved it. Then came the catnip and the peppermint... she decided that she would take her pinches of all the herbs, combine them, rub them in the palm of her hands and make her own concoction... and it was lovely.... Well, of course when daddy picked her up, she had to let him know what each plant was, have him rub/roll in his own hands and sniff... my budding little herbalist... and yes, she especially loved the eucakisses.

Escape

This morning started out early with the work here on the homestead. The air felt like mid-August although it is only late May… I won't be complaining though. Chris (my sister) showed up at the house at 6:30 along with her granddaughter, Jilly Bean. We planned on finishing another outside enclosure for my growing number of prisoners. As we worked on the enclosure, Daisy, Willow, and the black flies which needed their morning breakfast of human blood, crowded us…. Daisy, the nosy little goatee girl that she is, of course, had to affectionately climb all over Chris while she is twisting like a pretzel to hold boards while I screw them in… it is a serious wonder we didn't lose any fingers….

Finishing up in quick fashion (Oh, how we hurried due to being eaten alive by the NH State Bird), we moved to the backyard to finish securing the new fencing we had put up the week before… and of course, this time the hammer met my thumb… and good God of Homesteading I let out the worst of un-lady like words to ever pass these lips… well, maybe not ever….

Thinking we were done, Chris, and I moved to the front yard to rest… and what do we see? The Little Chicken Mafia converging on the garden I had spent all week planting… and Betty White, the little hussy, was digging up several of my potato hills… I threatened her with the stew pot… no solitary for her… roasted chicken was on my menu… so more fencing to run, more saplings to cut for fence poles, more NH State birds to battle… it never is easy out here on the homestead….

Finally, finally we were done... or so we thought….

And, the Little Nasties decided to make a jail break… so glistening (women do not sweat), bug bit and miffed off, I went running through the woods, batting aside pucker brush, getting scraped and scratched by tree limbs and low-hanging branches… all in search of those fluffy little fluckers… Chris running up the road searching underneath every bush and plant… Jilly Bean standing in the driveway, not knowing who to yell to first….

Well the fat little fluckers didn't manage to get too far… only to the nearest swamp about 500 yards or so into the woods… muddy, stinking, quacking, they waddled their fluffy little arses back to the pen… me walking shot gun behind them threatening them every step of the way….

And more fencing to fix…. Screw this fencing crap! I am beginning to hate fencing….

And they escaped again…. I had serious thoughts of roasted Nasties for dinner.…

Finally, settling down the prisoners I decided today would be a good time to plant Sunflowers a friend had gifted me… why not, I was already glistening and glowing, if I shined any brighter, I'd blind the sun….

No sooner had I got all 10 sunflower seedlings planted when I turned and behind me the Little Chicken Mafia looking like the feathery train from hell hot on my heels were digging up the sprouts and plucking the heads off as fast as I had planted them… Screaming I chased those little red-heads to and fro… threatening, screaming, face beet red and head about popping off… thank you God they can run faster than me… I swear to all that is holy, they will be on my plate sooner than later….

And more fencing… this time around all the garden beds.... Around all the elderberry, blueberry, strawberry, rhubarb, sunflowers, around every herb bed… and I am glistening... glistening… glistening….

Repeat Offender

As many of you know, the rooster, aka

Peckerhead escaped his cell this morning. I don't quite know how it happened other than the attendant, ie: me, was not paying attention and I will bet my last dime that he had been planning and plotting this prison break all night….

So, when I opened the cell door, baton ready in case he decided to gut me with a shiv, he squawked loudly, let out a growl and ran his fluffy, flucking little arse right out the cell door, right between my legs….

I dove for the tail feathers on the wily bad boy but he was faster than me… and he hissed and clucked laughter as he pranced and danced just beyond my reach… I threw his breakfast at him, marched into the house indignant that this ballsy rooster pulled one over on me.

Throughout the morning I sat in the dreary, cold waiting for him to come close enough so that I could pounce on him… catlike reflexes and God I so wished I had a taser… I'd of zapped his fuzzy little arse… and this gnarly little dude… wow… he taunted and teased, pranced and shook his fluffy little butt at me… daring me to give chase.

I finally caught my chance when he relaxed his teasing and went to check out the Little Chicken Mafia who had been calling from their pen enticing this bad boy over so they could flirt and fluff their own red feathers at him… his horniness was his downfall and my lucky break.

Sneaking up behind him made a flying tackle that would have made Tom Brady proud… and landed square on top of this flucking fat dude…. He let out a hiss and an indignant squawk as I wrestled him, pinning his claws and beak as he wildly tried to claw and peck his way out of my grasp. I rolled in the mud, him on top, me on top, him on bottom screeching, me on bottom struggling to keep a hold on him… he fought a valiant battle and I give him kudo's… he almost had me... but not quite good enough….

I placed him back into solitary. He pecked at my boot. I bopped him on his beak. He shook his head and gave me a startled look. He pecked at my boot again; I bopped him on his beak again… we did this several times… him not relenting and inch… and me determined to stand my ground.

"Look little dude… I can do this all day if I have to." I hissed. He growled in response… I know he was telling me in no uncertain terms to Fudge Off… Oh, the language this bad boy uses… shocking!

But as all stories go… and as all battles are finished, there is only one winner… and this chic rocks this barnyard! Prisoner Peckerhead = 0, Warden Nancy = 1….

Martial Law

The truce talks have been fruitless, and war has been declared once again… riots breaking out, burning, pillaging, total shit hit the fan situation. And General Peckerhead is leading the army. As Commander and Chief of the country, I have no choice but to declare Martial Law. It started 3 days ago when it was discussed that General Peckerhead would be allowed back into general population. His sentence had been agreed upon to be commuted for good behavior. Now the new prisoners of war, not three months in, were a fluffy gentle bunch. Minding their own business, following the laws set forth in the land, immigrants from another flock that had assimilated peacefully.

Day One: General Peckerhead's release went off without a hitch although it was witnessed by this officer that he was seen to be almost too friendly and chatty with the prisoners.

Day Two: As the officer entered the compound, General Peckerhead growled and fluffed his flucky little feathers, backing to the left quadrant of the compound. The fluffy flock followed him and there they convened a meeting that this Officer witnessed.

Day Three: Officer entered the compound for the usual morning duties. Witnessed General Peckerhead chase and demonstrate cruel and unusual punishment upon one of the fluffy flock. Officer then proceeded to none too gently kick the offending prisoner off of the victim. It was then that General Peckerhead turned and growled at the Officer. Advancing in a threatening manner. From the corner of her eye, the officer witnessed the gentle, fluffy little flock form a group behind the General… beady eyes trained on the Officer, hisses and squawks, insults and threat ensued….

The officer felt her life was in jeopardy… and she began to back away… falling into a kiddie pool filled with hay and landing none too gently on her ass….

General Peckerhead chose that moment to strike, flying at the officer, beak, claws and wings used as a weapon. The officer caught him mid-air and tossed him away. Jumping up, she chased him into the inner compound, the fluffy little flock hot on her ass… and she cornered him. He growled and hissed… puffed his flucking feathers up and made the charge. The officer had no choice but the give the General the business end of her steel-toed boot…. but he was tenacious… coming back with even more weaponry aimed at the officer. In her defense, she grabbed him up by his neck… and held on tight as his claws raked at her arms, leaving them a bloody mess… all the while the officer screaming for a lock down of the compound.

This is how war starts. Peace talks erode, chickens and roosters attack gangster style and the officers are over run….

But not this officer… She has restored a fragile peace… locking General Peckerhead back into solitary confinement.

I am happy to report that the compound is in lock down. Peace has been restored for the time being.

Blue Fuzzy Bird

Okay, it's been awhile since I've regaled all of you with homestead happenings... so I feel it is my duty to catch you up to date.

So, after spending the day cooking, canning and watching grand babies, I decided to cook a nice dinner for hubby and me. So I pulled out a couple of thick cut slices of beef roast that had been marinating in the fridge for two days, threw them in the iron skillet, peeled potatoes for mashing, sautéed some fresh picked sweet pea pods in the wok and had a mighty fine dinner ready as hubby walked in the door. Now, after the day

I'd had, I decided my jammies were in order early along with my Eeyore bathrobe…. Hey! Don't be haters… I like being comfy.

So, hubby and I are sitting eating our dinner he, in good sport, told me how great it was. And for the most part, it was good. Except for the meat... which was tougher than chewing on a leather belt. But my hubby being who he is, wouldn't complain ever about this ruined dinner. He's a champ like that. As dinner progressed hubby looks at me and the following conversation takes place.

Him: "What is that in the tree out there?"

Me: Squinting my eyes and looking out the window. "I don't see anything."

Him: "On the branch. I think it's one of those predator birds."

Me: Squinting harder. I can see a blob in the tree.

Him: "Your chickens are out."

Me: "Shit!"

I jump up from the table and run out the door. Not having anything on hand to throw at the damn bird which now I recognize as a chicken hawk, I start flapping my arms and squawking like some demented pigeon, blue bathrobe flying out behind me, flip flops on my feet, hair wild and tangled from just taking it down out of the braid it's been in all day… trying to appear larger and more looming… trying to intimidate the freaking chicken hawk. Watching it fly away… I'm thinking, yeah, I'd fly away too… fast, very fast if I saw this crazy Eeyore feathered bird squawking and running at me too.

I make my way back into the house. I see hubby sitting at the kitchen table, face beet red and tears running down his cheeks… first thought was 'Oh, my God I've killed him! He's choking to death on the meat!' I run over and yell.

"Are you choking?"

He shakes his head and snorts. Rather than choking on the tough meat, he is in fact choking with laughter. Indignant I plop rather ungracefully down on the chair and resume eating my dinner. After he stops snorting and laughing, he turns to me.

"I really think there is something the matter with you."

Me: "Fudge you, asshat!"

Him: "You looked like a big blue fuzzy bird out there squawking and yelling at that stupid chicken hawk… bwahahahahahahahha!"

Me: "I'm gonna punch you."

Him: Snicker snicker….

And this folks is how the day ended… with me throwing the dinner in the trash and hubby still chuckling on his way into bed… but my Little Chicken Mafia… well they were safe from that big bad chicken hawk… because this fuzzy blue bird scared it away.

Peace

There is a contentment on the homestead this morning. Perhaps it's because my canning season is in full swing, perhaps it's the animals or the gardens… or it could be just all of this. The two goatie boys, Rebel and Jasper, have made it their mission to call to me all day… their little voices ringing out from the barnyard.

Daisy and Willow, my two goatie girls I think are getting quite jealous as every time I enter their pens… I get mauled, stepped on, molested in true goat fashion. Now Peckerhead, the Bad Boy of the Barnyard, has a new flock and I wonder why he hasn't terrorized them yet. It might be that they are as big as he is? Or that there are two young roosters, Brutus and Bully, which he now has to contend with. I watch as they eye each other daily, flucking around in circles, growling and hissing… in other words… they are going to rumble very soon… and I think this time my bad boy Peckerhead might meet his match.

My kitchen looks like a production line for canning… jars stacked on the table and every available countertop surface. I am hoping hubby will get the hint and build me the shelving I've been asking for… for over a year now. Ahhh such is the life of living in a small log cabin.

The Little Chicken Mafia have been dancing around the front yard, given that they have been granted daily freedom from their coop… bug control at its finest. I have a cord of wood drying in a pile and beckoning me to get started stacking. But I just don't want to think about fall being right around the corner. I am enjoying the summer too much. And with the fall comes colder weather, then ice and snow shortly after and this thought just gives me chills. I am not ready to face that truth just yet.

So today I will concentrate on this contentedness. I will tend to my garden, tend to the prisoners in good, responsible warden fashion and I will pick from my garden. And my reward at the end of this day will be a good tiredness, a good book as I curl up on the couch and I am sure, and many good laughs as I stumble not so gracefully through the constant poop patrol, barn cleaning and animal antics.

Some Days It Just Doesn't Pay

There are just those days when you wonder if you should have just stayed in bed. And this just happens to be one of those days. My morning started out normal enough for my life here on the homestead. Watering and feeding the prisoners. Doing my daily poop patrol… things didn't really start falling apart until a few hours after breakfast. Today was hay day for me. So, my errand for this day was to make my way into town to the feed store and pick up hay for the goats and feed for the chickens. I do this once a week and normally it is no problem.

But of course….

I always seem to run into the problems… like a bad apple in a bag, I manage to find it. Order of business was to put the 4 mischievous goats away into the barn while I ran my errands. If you know anything about goats, you know you never leave them unattended. They always will find trouble. So, grabbing a handful of cookies, I made my way to the pen and coaxed them into following me into the barn. The fat little critters love their cookies. Opening the barn door slightly, I turned to grab the water hose to refresh their water and had my hand on the frame of the door. It was then that Daisy, a sixty plus pound butterball of a goat decided to slam her full body against the door… effectively trapping my hand. Excited about her cookies, she pushed her weight onto the door and wouldn't budge… and her sister, Willow, excited for her cookie too, joined her. So now I have over 150 pounds between the two of them playing blockade on the door where my hand is smooshed… and bleeding… and I am yelling to high hell and pushing back… with my little 110 pounds of pissed off fury….

Next on the agenda after cleaning up the blood and bandaging my hand was to go downtown for the hay… now normally when I am working on the homestead, I am not dressed in any fashionista style… in fact, I probably look more like a homeless bag lady. The weather here in NH has been triple H… hazy, hot, and humid… so I had only dressed in an undershirt, shorts, socks, and sneakers. Not my going to town outfit by any means. In my haste, pain and frustration I jumped into my truck. Over the weekend hubby changed the back brakes, so I figured it was good to go right? Wrong! Driving down the road I came to a hill and tapped the brakes to slow down… and I heard the most God awful sound… and the truck started jumping like a jack rabbit on steroids… and I am now wondering if I am going to die….

Yelling and swearing... threatening hubby in every possible painful way I can think of, I proceed to the store. (He is at work but boy when he gets home!) Fudge it! I would ride it until it either fixed itself, burned itself up or died.

I make it to the store, and it was then it dawned on me….

I was not dressed appropriately to be anywhere near the public. You see, I am a bit modest when it comes to wandering around in public… I would normally wear a bra, a nice shirt, jeans and boots… and today, guess what, an undershirt with no bra… Oh, I am mortified by this. Gritting my teeth and praying that I would not see anyone I knew… face red with embarrassment, I walked into the store which of course was crowded with EVERYONE I knew…. I tell ya my friends; I wanted to crawl under a rock and hide.

As quick as I could I paid for my hay and drove the truck… (with now a stinky smell coming from the wheels) around to the back of the building so a nice young man could load the hay… and of course, there were stacks and stacks of pallets in front of where I needed to be… and of course backing in between these stacks I just had to hit one with the bumper of my truck… and like dominoes… they fell… the whole fudging… pain in the arse stack… falling… falling… falling….

The warehouse dude looked at me… I looked at him… he loaded my hay without saying a word…. Thank You God!

Pulling out of the back entrance, my only thought was to get the fudge home… wrap myself in a good book and hide away. But, nope, it gets even better! You see, sometimes the universe loves to challenge you and today… she was throwing every possible obstacle she could think of in my way.

And it just figured… there had to be a tractor trailer unloading. And this dude left very little space between the building and his truck for me to get by. So, inch by inch I maneuvered the truck down the narrow space… and... I will be damned if that truck driving dude did not step out in front of me and start yelling to high heaven.

Okay, so I am a fairly patient person. I try to do my best to be kind and considerate and courteous but there is only so much a person can take and I had reached my limit….

Some dude, standing in front of my truck, blocking my path… yelling at me in a language I don't even understand… whelp? I don't remember much after that….

I jumped out of the truck.

Him, waving his arms and dancing around: "Blah, blah, blah… fudge you!"

Me: "The fudge you say?"

Him pointing to me, my truck, his truck.: "Blah, blah, blah, fudge you!"

Me: "You want me to what? Show you how to park that thing?"

Now this dude is seriously mad, and I am getting hotter by the second…

Him: "Fudge you!"

Me: "Oh, no you just did not! Not unless you put a ring on this finger and even then, your chances are slim to none dude!"

Him throwing up his arms and walking away. Well, that made me even more determined to piss him off, so I follow him.

Me: "Dude! What is your problem?"

Him: "Fudge you."

I wonder if he understood English?

Well, happy ending to this story is that I made it home. Now I will spend the rest of the day unloading hay, cleaning the barn and plotting my husband's slow and painful demise….

Hornets…Oh, My

Well today was one of those days you either want to forever forget or chalk up to lessons learned. Hubby is on vacation, so one of our projects to work on was the bedroom. Last winter we dang near froze to death in the bedroom. We plastic the window, we plugged holes with insulation, yet we just could not keep the drafts down. Some mornings it would be so cold we'd wake up and see our breath. I was not impressed, and I won't mind telling you, I bitched all winter long about it.

So, first thing after sunup I started in as the wrecking crew. Hubby says I am wicked at demolition. And I am. It feels good to rip and tear, and smash… I really do like it… and it's a great way to take my aggression out. So, I am ripping paneling off the wall to find insulation that is disgustingly smelling of mice pee and poop. Yup, you know all about the mice in my house, I've developed quite the relationship with those little fuzzy poop machines. As I am pulling down this really gross insulation and mind you, gagging, I hear a buzzing. In fact, lots of buzzing. Pulling away sheet after sheet I am exposing one hornets' nest after another, two in the corner near the ceiling, two more on underneath a cross brace, hell more here, more there to find a total of ten nests in my wall. And because I am disturbing them, they are leaving their nests, in one angry, flying, buzzing horde! Hubby yells, I yell. He's swatting, I grab the hammer and start swinging… knocking those little flying stinging bastards right out of the air… hubby in bopping, ducking, and weaving as I flail about with the hammer chasing hornets… Hubby is yelling, "What the fudge! Put down the fudging hammer!" And I am hell bent on whacking each one of those buggers.

Finally through my murderous haze I hear him… and I drop the hammer and grab the vacuum… if he won't let me splat them with the hammer I will suck those fudgers out of the air… and I did ... and their nests… all into my $99 Power Force Vacuum with a built in canister that I can see through… and I see some very pissed off hornets… and this pleases me.

Did I say I hate working on construction projects with my hubby? Well, he took my hammer away and made me instead work with the staple gun… I think he thought it would be safer….

Little does he know; staple guns can be a very dangerous weapon should the right occasion arise….

How Many Showers Does a Homesteader Take?

Well, this morning I set my feet on the floor, ignored the eruption on my face that is threating to send my head into orbit and decided the day was going to be productive. Well you know what they say about the best laid plans, right? The critters were crying, Peckerhead letting out his usual 4 am crow so I strapped on my snow boots and made my way out to the barn… and was met with some very cantankerous beasts…. The little billy, Jasper decided first off he was going to jump on my back and start humping me…the horny little turd…kicking him off I came face to face with Peckerhead who, despite his ass kicking yesterday, wanted another go at me. Now mind you, it's raining, it's icy, and I am trying to load in hay, fill grain dishes, carry water… and am soaked through three layers of clothing….

Avoiding an all-out bloody battle with Peckerhead, I finished up my barn chores, settled everyone down and then realized I still had yet to fill the wood box and clean the soggy, poopy hay from the animal stalls….

So, grabbing my best girl (Tara-belle) we set to work. Now the weather is beautiful in spite of the heavy rain. The air is warm, and the ground is sloshy. Terra-belle is loving it and so am I, that is until she wipes my feet out from under me landing me in a soggy, shitty mess…a pile I had shoveled from the barn is now up my back, under my shirt, dripping down the back of my pants…I've got two nosy, affectionate goat girls standing over me, a Peckerhead ready to pounce and a sixty pound dog frolicking like it's play time… and Oh, the smell…. Gagging I pulled myself to my feet, yelled at my dog, pushed the two bully goats away and made my way into the house to strip and shower…all before nine a.m….

And this my friends is how my day has started.

I'd Rather Battle Peckerhead

Every once in a while, a monkey wrench gets thrown into the life of this homesteader and I swear it's the universe just trying to keep me challenged. Earlier in the week I had an episode with vertigo… no big deal, I've dealt with this many times… but this time I lost my footing and rammed my face into the door sill….

On my right cheek I have a cyst, it's been there for quite a while, not bad but rather just an annoyance. When I fell into the door sill, I hit my cheek and thus ruptured the cyst. By Monday night, my cheek had swollen to the size of a big marble…hot, red and tender to the touch. So, I called my doc and scheduled an appointment thinking she'd just throw an antibiotic at me and call it good. But the universe had other ideas. By Friday, my face was swollen, angry red and Oh, so sore….

Doc: "Ouch, that looks nasty."

Me: "Ya think?"

Doc: "I ain't touching that with a ten-foot pole."

Me: "What'cha mean?"

Doc: "Nope, you're seeing a surgeon this afternoon girl."

Me: "Oh, ffs….can't you just slice it, drain it, stitch it up and send me home with antibiotics?"

Doc: "Ummmm, I don't think so."

So, off to the surgeons I go, bitching all the way that she had better lance this bitch and send me home.

Surgeon: "Oh, that looks nasty. I bet it hurts."

Me: "Nope, walk in the park lady, doesn't hurt a

bit."

Surgeon: "Well, let's lance it and see if we can get that bad boy to drain

Me: "Yup…do it!"

She proceeded to drive four needles into my cheek about sending my brain into orbit….then she pulls out her scalpel with glee and made the first cut….and I damn near came up off the table grabbing her wrist….

Me: "Fudge, did you numb it?"

Surgeon: "Yes, but it's abscessed so it probably didn't take."

Me: "Then lady you better hit it again with more medicine before you decide to slice into me again!"

Surgeon: Rolls her eyes. (At this point I'm thinking she's enjoying inflicting pain on me)

She pulls out a needle and sticks me another four times, each time I'm gritting my teeth against the pain and envisioning driving the needle right into her left eye…

Surgeon: "Let's try this again." (as she comes at me with the scalpel)….

As she is digging, slicing, dicing, pinching and pulling I am muttering threats of death to her… and she just smiles at her assistant like she's heard it all before. Little does she know I am very serious about doing her bodily harm.

Surgeon: "Ummm this is too deep. I'm not thinking this is a good idea."

Me coming up off the table furious.

"What is not a good idea?" I yell as blood is trickling into my ear and my eyes are about to pop out of their sockets from the pain this scalpel wielding woman is inflicting on my poor face.

Surgeon: "The abscess and cyst are deeper than I thought. I can't get to it to remove it. It's too, ummm, let's say chunky? So, here's what we're gonna do. I'm gonna place a drain in the wound, give you antibiotics and we're gonna have to schedule another surgery on Monday."

Me: "Oh, I so hate you right now. And what the hell does chunky even mean?"

Surgeon: "See you Monday morning. We'll get this fixed right up."

Glaring at her I nodded.

So, I am sitting home on my couch, warm compress on my face, my right eye about swollen shut and thinking this is just another day in the life of this homesteader… monkey wrench and all. What the fudge!

In the Jungle the Mighty Jungle

Well today I had my own SHTF situation on the homestead. There was plenty of swearing, there were weapons of my destruction, there was blood and kicking and screaming and yes, even prayers….

I went out to do my usual morning poop patrol and farm chores. Gathering up the buckets of feed, water and hay, I first cuddled with the babies, gave mama Daisy a bit of extra love and then off to the barn to tend to the other critters. Now the Little Nasties (my ducks) were quacking and raising their usual ruckus demanding their breakfast so tiptoeing to avoid the slime of duck poop, I tended them first, to frankly just shut the fluffy little fluckers up. Then into the barn with my buck boy and his brother, my pregnant doe and of course, my nemesis, Peckerhead. Now Peckerhead earned his title, many times over. He is one cantankerous rooster. Most times we just warily circle each other. He growls and hisses at me and I threaten to turn him into to rooster stew. Only occasionally does he follow up with his threats though, so I wasn't prepared this morning for his Rambo style attack.

Walking into the barn I have my arms full. I open the door to buckie boy pushing his fat arse against my legs trying to dislodge the bucket of grain, of my pregnant mama tangling around my feet like a cat, tripping me up and of Peckerhead up on his perch strangely quiet and staring down at me. I push the animals out of the way so I can fill the hay bin and their dishes. As I am bending over, by the way not a great idea with a buck nor a bad ass rooster, I hear a growl and a hiss.

"What the fudge dude!" I snapped at Peckerhead as I stand up. He is face to face with me and I back up a step. First mistake… I know better than to show weakness and backing up, to him, was a clear sign of weakness.

"Don't even think about it!" I growl back at him. Now the tone of my voice, the threatening posture of my body must've challenged him because with an indignant squawk, he launched himself at me. Throwing my hands up, I stumbled backwards, tripped over the water pail and the Buckie boy and landed flat on my arse in a pile of stinky, wet, poopy hay… and this

Rambo style attack is an all-out blood lust war. Peckerhead was on me, claws, beak, spurs and wings… like a mini pterodactyl stabbing at me as I punched and kicked to get the fluffy little flucker away. Growling and hissing, me not him, I landed a right hook to his head, knocking him back. Buckie boy is trying to get out of the fray and takes a flying leap like a gazelle and about dang near takes my head off with his hooves as he sails over me. Ducking and scrambling through poop and slime, I manage to make it to my feet as the Rambo attack continues and now I am so pissed off that I could easily wring his neck should I get a hold of that cantankerous rooster.

Standing, I am breathing heavy and my heart is chugging like a racehorse who had just done laps. Peckerhead is tangled in my legs and spurring the shit out'a me and looking for blood. I'm kicking and swearing….

"You little fluffy motherfudger! I will end you!" I scream as I turn the tables and take after him. He backs up, scoots around me and comes in from behind. I spin around like a drunken ballerina and catch him midair, wrapping one hand around his scrawny throat and squeezing as he beats his wings against me like some rabid bat stretching to get to my throat. His eyes are glazed with hatred and blood lust and this little fluffy flucker ain't giving an ounce. I thought today might actually be the day I have to finally end our two-year relationship. Frankly it hasn't been going well at all and I don't think we'll ever make it to the proverbial alter. It's him, not me. I wanted to love this little man-eating pterodactyl, but he is making it so hard. Sneak attacks when I least expect it, blood lust, he has no respect for his human. I just don't think it's gonna work out. Long term relationships require effort from both parties and I just don't see him giving much effort toward anything but trying to kill me.

My ass covered with poop and hay, my Buckie boy cringing in the corner of the barn and Peckerhead finally giving up the fight, I made my way back into the house to clean up and tend to some pretty knarly scratches. Today he will learn a hard lesson in love. The fluffy little flucker with the bad attitude will be my constant companion.... he is sentenced to spend the day under my arm, carrying him like a football, as I work around the homestead. He will squawk and protest, he will growl and hiss, but he will once again learn to love me....

So maybe, just maybe, perhaps, this relationship thingy still has some hope.

Easy Peasy Right?

Well, another building day on the homestead and hubby has the man-flu, so he played supervisor while I played the grunt worker... or the builder. Needless to say, we got the new stall built for my pregnant doe, but not without plenty of swearing, a little blood and a thrown hammer or two... or three... or five times.

Hubby: "You are not a good team player."

Me: "Fudge you."

Hubby: "You need to screw this board to the floor at this corner."

Me: "I'm starting in the middle then I'll work

my way to either end."

Hubby: "You are supposed to go from one end to the other."

Me: "Why? What fudging law says I have to?"

Hubby: "Do what you want."

Me: "Always do."

Meanwhile, Daisy mama is chewing on my hair, sticking her tongue in my ear, jumping on my back as I am kneeling down on the floor, knocking over my tools and the two goatee babies are frolicking all around my feet, my hands, my legs… I was in the midst of a troop of chimpanzee's hell bent on making my work so much harder. And hubby is laughing.

Me: "What's so funny?" (As I am cleaning up blood from a finger, I just drove a screw into).

Hubby: "How can you get anything done with these critters crawling, jumping, and nibbling on you?"

Me: "Same way I do with you? I just dig down and get it done!"

Hubby: Shaking his head.

Me: "Teehee."

Goats are screaming, hubby is Supervising, I am fighting with tools, mama goat, baby goats and bloody fingers and the biggest baby of all who has the man-flu.

But, the fudging birthing stall has been built and now my fat pregnant mama can have her babies….

Such is life on this homestead… always a challenge.

Night of the Bat

Now my life here on the homestead has had its ups and downs, and the stories I've written over the past two years have become a collection that I will someday turn into a book for my kids....

This story, I share with you from two years ago. I dislike creepy crawly things... go figure right? As a homesteader you'd think I'd be well past that by now... but no....

The shits and giggles just never seem to end for this tired homesteader and last night the party lasted until well after midnight.

After a long day which started with the usual poop patrol, playing referee to the Little Chicken Mafia and the Bad Boys of the Barnyard, running chicken wire fencing over the top of the duck coop that left my arms scratched and bloody I was finally looking forward to a quiet and settled evening of relaxation. After dinner was served and cleaned up, I grabbed a book and sank deep into the softness of the couch. Ahhhhh... sweet silence.

Then the unthinkable happened.... Hearing a thump, I decided to ignore it thinking it was just my curious, rambunctious kitty gearing up for her nightly antics of chasing toys and banging cupboard doors. Generally, just being an annoyance until I finally would yell at her to stop already!

Back to the original thump though. This was not the usual thump because this thump turned into a series of squeaks and clicks....

Holy Mother of God Why? Why was there a black shadow on my floor, inches away from my foot surrounded by three dogs who were now circling it like a pack of hungry hyenas....????

Putting my book down I stood up and took a closer look then jumped back as a scream erupted from my lips.... "Flucking Bat!" I yelled. I picked up my husband's slipper and threw it as hard as I could at this leathery, scrabbling little flying rat bastard.

Now I forgot that my dog Tara loves my husband's slippers and when I threw it she dove and gave chase which in turn caused the offending little creature on my floor to explode upward with a flurry and flutter of wings and swoop loop-de-loops over my head. Diving, I hit the floor and crawled under the kitchen table with three dogs on my ass and one of them I think trying to hump my foot... that or crawl under me for protection. I've yet to figure that one out.

As I hid, no really cowered, under the kitchen table that flying rat bastard swooped and dived bomb us from above... all the while squeaking and chittering angrily....

I realized that I had only two choices in this situation.... Either spend the night under the kitchen table or get up and chase that little flucker out of my house. Crawling out from under the table, I grabbed the couch cushion and began flailing wildly to knock the offensive little flying mouse out of the air. I didn't have a prayer in hell of capturing it but I did have one lucky swing and managed to hit it dead on and knock it to the floor where I quickly grabbed the nearest dog dish and effectively trapped it beneath it.

With shaking hands, I grabbed a manila envelope from the end table and slid it between the floor and the dish, all the while praying to any deity that would listen to please, please, please not let this thing escape.

Carefully I picked up the envelope, the dish and the bat and carried them out to my front yard where I placed the whole set up down on the grass. Then I went and grabbed a flashlight from the hallway and went to search for the longest stick I could find. Mind you I am in my blue and white Eeyore bathrobe; a pair of flip flops and I thank God I don't have close neighbors!!!!

What a spectacle I must've looked. But back to the stick.... I wanted it to be able to put as much distance between me and the flying rat as possible when I lifted that dish.

Cautiously I reached out with the stick and lifted the dish off of the bat. It lay on the envelope absolutely still. I poked it gently with my stick... still no movement. Gosh I hope I didn't kill the little thing. Sucking in a deep breath I moved a little closer and poked it again and this time it sent out a series of angry little chitters. Okay, whew, I hadn't killed it. Moving up onto the porch I sat and watched, hoping it would take wing and fly away.... And also hoping it wasn't pissed off enough at me to fly toward my face and attach itself like that octopussy creature in Alien had done to Sigorny Weaver....

After about ten minutes of watching I finally got up and decided that it would either fly away or be food for one of the night hunters that often roamed my yard. Either way, I was tired and going to bed. "What was all the ruckus?" my husband mumbled sleepily as I crawled in bed beside him.

REALLY? WHO THE FUDGE DOESN'T WAKE UP WHEN THEIR WIFE IS SCREAMING TO BEAT ALL HELL IN THE MIDDLE OF THE NIGHT?

I secretly and quietly entertained visions of

stuffing a pillow over his face and suffocating him. Rat Bastard Be Gone!

A Weapon and A Woman

Well, Peckerhead was in the foulest of foul moods today.... I wonder what feather got across his ass....

My day started out early, feeding and watering the critters, sending them all to their outdoor pens to enjoy the sunshine and booting the big Mister Peckerhead out of the barn to free range and patrol the perimeter. With poop patrol done, I set off to clean some brush, break it down to more manageable pieces for my burn pile later next week. I got a new (well used but new to me) Nano saw and the thing worked slicker than greased pig on a pole.

I'm sawing away on the saplings that I'd cut earlier in the season... just scrubby annoying trees that were clogging up the edge of the driveway and vomiting branches with every storm. I look behind me to see that fat little flucker in stealth mode, creeping up behind me as I bent over the logs. Turning, I warned him in no uncertain terms that I'd kick his fluffy arse if he came at me. His beady little raptor eyes stared, coldly, with a deadly glare right into mine.

"Nope, don't you do it!" I yelled and threw a stone at him .... he squawked and skittered away. Thinking I was safe, I turned back to my work only to feel a sharp peck at my leg ....

"Oh, you little flucker!" I yelled as he's beating feet in a full out retreat. Picking up a stick I gave chase and was gonna beat the tar out'a that dinorooster… when he flew up onto a nearby branch and roosted, staring down at me. I threw my arms up in defeat and listened to his laughter as I made my way back to the task at hand....

Then he struck... and I yelled, I booted him with my steel toed work boot, he squawked, ran straight at me like the demon spawn he is... then right past me as I swung down my nano saw damn near slicing my leg in two... and he kept running... and squawking... and feathers ruffled up... full on assault mode.... I had dislodged a mouse from my brush pile... and the little flucker wasn't trying to attack me, but rather, he was trying to fill his fat little belly with fresh mouse buffet....

Now, I feel really horrible that I tried to cut that little flucker in half with my new nano saw... but hey, in my defense, I really thought it was gonna be another all-out brawl between the two of us.... Oooops.... Peace out peeps and enjoy your day .... ya never know when you're gonna be eaten by a dinorooster....

Karma

Here peeps. Hold my iced coffee.... Well, today didn't turn out quite as planned. First item on my to-do list was to start building a new shelter for my ducks and to clean out their existing shelter for 10 meat birds I ordered due to arrive mid May... soooooo.

After my usual morning poop patrol, feeding and watering the critters, I set my tired ass off to start my building project. I also had brush to burn so fired up the pit and got a blaze going. Hauling out the skill saw, the hammer, the squary thing for drawing straight lines to cut, the carpenters pencil, my measuring tape, I set off to the reclaimed wood I got from my sidekick sista's demolition pile and began to pull nails and unscrew screws....

Now, Peckerhead was kicked out of the barn at dawn as he is not allowed to mingle with the ladies... he's too brutish... and I really need to contact the Femnazis and Vaghats to arrange a protest of his bad behavior... but that's another story for another day.... I'm fighting with the nails that, of course, are driven in and bent and lo-and-behold I got a puncture to the palm... just a little one, only bled a drop or two....

Then I couldn't for the life of me, figure my dang measurements so had to jump on fb and give a shout out to the carpenters for some help and after a few chuckles, a few suggestions, I think I got it figured out. So, pulling nails, unscrewing screws, measuring boards, cutting boards and tending my fire....

All the while my Peckerhead is at the fence line bitching and moaning because he's getting wet... yes, I am doing this project in the drizzly rain... covered in mud, listening to a pissed of rooster who is taunting me with all manner of threats... and posturing's....

I raise my hammer as he comes toward me... I back him off... with a big board. He wanders back to the fence line. I go back to work. Then it starts to pour... and I gather up all my tools and in the drenching downpour scatter and crowd the ladies into the barn.... Peckerhead is screeching at me... he is muddy and filthy and so embarrassed. I am laughing at him.... He is pissed off....

I am enjoying his discomfort a bit too much as I watch him pace back and forth in the downpour.... (No, do not call the insane Peta peeps on me... the fudging rooster has shelter... he was just too stubborn to go under it.)

Finally, I take pity on his poor fluffled feathery ass, scoop him up in my arms and duck walk him to the barn as he is growling and hissing and struggling to attack me....

I am stronger and he doesn't even get a peck in... lol.

Touché Peckerhead. Karma is a bitch, and she is named Nancy!

Attitude

Now, hubby says he knows where Peckerhead gets his asshole attitude, and I gotta admit, I think he's right... lol.

Toasters! It drives my hubby crazy that I won't buy a toaster... but my reasoning is, why should I? The oven works just fine for toasting bread... right?

Yesterday, toasting my bread, I set the smoke alarm off. What can I say, I'm a multitasker with shit for a memory….

Hubby: "Buy a toaster."

Me: "I don't need a toaster."

Hubby: "I'm buying you a toaster."

Me: "I'll just give it away, besides that, if the shtf, then a toaster won't do us any good."

Hubby: "If the shtf, we won't have to worry about it, we won't have bread, anyway."

Me: grinning, "Honey, you know we'd have bread, if there's anything I'm good at, it's making bread...."

Hubby: "Just buy the damn toaster."

Me: "Nope."

Now, he did this with clocks a few years back... kept complaining that he never knew what time it was cuz the clock batteries were always dead. So, being me... I went out and purchased clocks, lots of them, lined them up on one wall, but... teehee... set each clock to a different time... poor bastard still didn't know what the hell time it was.... Bwhahahahhahahahhhhha.... Maybe today I should go out and buy ten toasters, zip tie the cords together and plug just one in... so he'll then have to figure out which one is live and which one is Memorex... teehee....

Yes, gotta yank on him every once in a while, to remind me who he's married to....

"Bwhahahahhahahahhhha!"

Decisions, Decisions

Well, snow, slippery under foot, does not bode well for my ass... first thing I found this morning were tracks outside my chicken house. Ummmm okay, so another little bastard will put my skinning knife to good use, off to feed the critters, scoop the poop, chip ice... my usual... until....

Evidently my Peckerhead was on alert due to said critter... and not in a good mood this morning.... In fact, he kinda reminds me of a cross between an ill-tempered toddler and a really gnarly pterodactyl....

Opening the coop door... we come eye to eye... him up on the roost, peering down at me.... I move ever so cautiously as I listen to his growls.... As I bend over, reaching for the eggs.... this mofo decides to pounce... sending me ass over tea kettle back out the door....

I explode up out'a the snow, swearing like sailor, swinging like prized boxer... he flies out'a the coop at me and I co-cock him right in the noggin, knocking him into the snow.... Well, a wet, snowy pissed off rooster is not what you want to mess with first thing in the morning....

He came back, met the business end of my boot. I kicked, he squawked, I dove for him, missed, got a face full of snow, he charged back, I ducked, threw my bucket of water at him....

He is now up in the tree, laughing down at me....

Mmmmmmmmm.... Handgun or shotgun... can't decide which.

Trouble and Temper

Wow... this lady is tired tonight! Today, started out early with my morning poop patrol and Peckerhead was in rare form... rubbing himself against my legs, I think he was purring... yes, purring.... So anywhoooooo... got my chores done, headed down the mountain to sidekick sista's where we started off canning 8 pints of Habanero Apricot Jelly, had no gloves so guess what? Yes, fingers are burned and stinging... washed hands thinking I washed the stinging oil off, welp... let's just say... DO NOT go the bathroom and grab the t-paper... owie.... There are places burning that just shouldn't be.... Then out to the demo pile where we spent a better part of the morning sweating, fighting the black flies, doing the NH wave and busting our arses... losing our footing, swearing... yeah, you get the idea.... Loaded up the boards I needed, headed back home to just unload all that I loaded, pulled nails, cut boards to size and got the whole pile stacked for attaching to the duck house this weekend. Peckerhead, all the while curiously watching me. Then into the house where I made 45 meatballs and spaghetti for dinner. Back out to put the critters in for the night, and he freakin struck! Heard the hens bitching... peeked in the coop to see that little flucker beating up on one of the girls. Furiously I stormed through the door screaming at that fat little bastard... he turned, launched himself at me.... I caught him mid-air and tossed him... he hit the wall... slid down with a furious squawk and growl... then dove in for a second go.

I slapped him across the head... he punched me with his feet.... I swung wide and missed.... He landed a good spur onto my boot.... I kicked... he flew.... I ducked and finally got my arms around him... and squeezed.... I thought his eyes were gonna pop out....

"Hah! You little jerk...."

He landed on the floor... yes, I had to let him go.... and slunk his loser little ass into the corner where he hissed, growled and hurled death threats at me... Are you kidding? He was just loving up on me this morning?

Yes, he is bi-polar. And either he is gonna need meds or I am.

Not sure which yet... bwahahahahahhahaha! Peace out peeps.

Temper, Temper

Well, never, ever, ever piss a grandmother off! Today, the sidekick sista and I were spending our morning out running errands and had lunch plans. Her treat for all the hard work I helped her with this week on her own little dirt patch. She's also been babysitting her granddaughter all week so lunch out was a special treat for us all.

We're sitting in Kimble's Ice-cream/Seafood having our lunch... it is the best place in all of NH for seafood and their reputation for excellence never fails... over stuffing ourselves, enjoying our time away, chit chatting and laughing.... We finish our meals and box up the leftovers. I carried them out to the car while sidekick sista was cleaning up the Little Jelly Bean.... As she's buckling her into her car seat, no easy feat with all the snaps and buckles they have, this lady starts to pull into the parking space beside us... back door of my car wide open, sista sidekick bent over trying to fight with the buckles on the not so adult friendly car seat... and the woman keeps inching forward... and forward....

I am looking in my rear-view mirror... and I hear the growl... sidekick sista is rushing trying to fight with the car seat... and this woman keeps inching forward....

I grin.... I know what the fudge is coming.... Sidekick sista: "Fudge lady, give me a moment."

Strange Lady: "Moves forward another inch or two."

Me: Starting to giggle.

Sidekick Sista: "Okay, you are now beginning to piss me off!"

Strange lady: Inching forward

Me: Laughing and trying to unbuckle my seat belt....

I know what coming and I'm trying to get out of my car to break up the brawl that is sure gonna happen.... You see, my sista has a temper that equals mine in force... and explosion!

Strange Lady: "Take your time why don't ya..." she yells out her window....

Me: "Ruh rOh...."

Sidekick sista explodes out of the back seat....

Me: Laughing and still fighting with my seat belt.

Sidekick sista: "You m'f'a'b, I'm gonna rip you out'a that m'f car you m'f' b' a.... beep, beep, beep, beep...."

Me: "BWahahahahhahahahahahhahhha.... OH, SHIT!"

Sidekick Sista as she takes an angry step toward the strange lady's car yelling: "You fudging crazy bit**** get your ass in this car now!" As I am still fighting with my seatbelt and laughing so hard I think I'm gonna have an accident....

Sidekick Sista whips around and looks me dead in the eye, "I think I can take her!"

Me: "You lunatic, crazy ass woman, of course you can take her, that's not what I'm worried about.... I don't have any money for your fudging bail! Bwahahahahahhahahahahhha."

Oh, my gosh... bwaahahahhahahahahahhhha! So, lesson for all you peeps out there... never, ever, ever, piss off a grandma.

Planning and Plotting

Well, the fluffy little flucker is at it again... lol. So, my little chicken Mc'Nuggets keep escaping their enclosure... fluffy little white asses traipsing all through my yard and garden... me chasing them at the asscrack of dawn... there ain't enough coffee on this planet this morning to keep me from thinking thoughts that no sane homesteader should be thinking. I'm trying to wrangle up these white puff balls and on the phone with my sidekick sista when she hears me yelling at the ornery roosters... "Peckerhead....!!!!!"

Me: "Don't you fudging do it! Get the fudge away from me or I swear you'll end up dumplings today."

Her: "What is going on?"

Me: "Peckerhead just fudging charged at me!"

Her: "Shoot him!"

Me: "I thought about that... maybe I'll get the pellet gun and put a few pellets in his fluffy arse."

Her: "Or buckshot?"

Me: "No, blow him apart."

Her: "How bout a paint ball gun, we could color him up nice and pretty."

Me... laughing: "That is not a bad idea!"

Her: "It would hurt him."

Me: "Well he hurts me all the time so...."

Her: "Yeah, but I kinda like that fluffy flucker."

Me: "Me too unless he's plotting to kill me."

So thus we are planning the peppering of my gnarly little asshat rooster.... God, this homesteading gig is a hoot sometimes....

Side Kick Sista

As I am sitting here on the couch, bemoaning my aching arse... another morning spent at my sidekick sista's helping her out with the enormous job of cleaning up the yard... now I say enormous because who ever lived there previously, welp? They were just human pigs... that's all I can say... and I say that with the utmost kindness....

Dilapidated buildings filled with garbage of every sort, a back forty that is nothing but debris pile after debris pile of rotten woods, brush, broken bottles, trash, heaps upon heaps of old tires, bicycles, kids toys and ya... you get the idea... a pig sty of epic proportions....

For the past two days we've tackled a brush pile of downed spindly trees that was the size of Mount Everest... and beneath it... lo-and-behold, trash.... Boards... bottles and more of the same... to make matters worse she has this vine growth that has tied everything up into one huge rats nest....

Hauling, lopping with loppers, sawing through branches... and Emily, her thousand pound German Shepherd, biting at the branches, biting at the rake, digging holes, carrying off tools... the fluffy little flucker doing her dangest to trip me up with every step... and God I love that damn dog but my patience will only stretch so far before I'm screaming and chasing her with the rake... which she thinks is all fun and games....

Until I landed on my ass... by stepping in one of the holes, she'd dug and tripping backwards over a stump.... And the sidekick sista....

Laughing so hard she's about ready to piddle her britches....

And my ass is covered with mud....

And she is pulling me up off the ground, tugging my tee-shirt up and trying to pull my pants down so she could see if I am cut/scraped....

And I am yelling at her to gtfo of me... get out'a my pants... all the while fighting her dog with my rake....

I would rather face Peckerhead than that Emily dog... and my sidekick sista!

Peace peeps... now I've got farm chores to do... let's see what this afternoon brings.

Bad Dog-Bad Rooster

Dang, I do not think there is a day when I'm not diving shit deep into one mess or another... and always; it includes that fluffy little flucker Peckerhead.... This morning I was busting my arse... pulling 2x4's so I could yank the nails and get ready to start my additions roof... so cruising smoothly along, I'm just doing my thing... then into the house to do laundry, make mac salad for dinner, clean, and all the other million and one things I get to do... teehee.... Then lunch, back out to the yard to start mowing....

Now I'll Tell ya, it's hotter than the flames of hell licking at my arse... and I'm sweaty, stinky, and at this point dang tired... lawn mower going, my pups are out... all three... two little ankle biters and my Tara belle... minding their own... until I hear a yowling over the sound of the mower....

Well I drop and run, cuz this type of yowling is usually not a good indication.... I see that fluffy little flucker going at it with one of the ankle biters… Sadie... and I scream....

"Peckerhead get the fluck off of my dog!" He turns, gives me a beady little stare... hisses and grins... yes, my flucking rooster grins and it is pure evil! And goes back at chasing my little dog.... So, Peckerhead chasing the dog, the dog running like all Hell's demons are after her; me chasing and screaming at the rooster....

Swearing, hot... my brain explodes and I make a flying dive... and miss... and that little flucker pounces on me....

It is war! I beat his ass so that his own mother wouldn't even recognize him.... I've got scratches on my arms, on my back, he pulled my hair, kicked, spit....

What a dirty little fighter he is....

I punched, rolled, pummeled, and that little flucker kept coming back for more.... The ankle biter was barking, Tara was jumping on me like this was all fun and games, the goats were screaming.... I am at the bottom of a virtual pile up... and it was all fun and games until someone got hurt... namely me.... Oh, my aching ass!

So Peckerhead is sulking under the porch, I am a walking wounded but ya know what... the fudging lawn got mowed! So take that you little rat bastard Peckerhead....!!!!

Chaos

Well, I knew it was too good to last. On these last few days Peckerhead has been unusually quiet and subdued. Out of his normal... so this morning, I'm at my usual slave labor, but with a twist... ice! Slipping and sliding my way along, busting ice out of the buckets, walking like a drunken penguin, I make my way to the barn... throw grain and hay at the goatees as they are horning me, jumping on me, doing their damnedest to take my feet out from under me....

Now Tara-bell, my doggie, goes everywhere with me, she's my little sidekick, protector, pal. I let the fluffy flucker out of the barn and shut the door so I don't have to worry about a sneak attack when my back is turned... those of you who own roosters will know very well what I am talking about.... They are little assassins... just waiting in the shadows to pounce...

claws, spurs, and beaks....

I hear a yipping outside, followed by a howl of pain and think my Tara is getting it on with a coyote or some other wild critter... so out the door I fly to protect my little girl... wipe out on the ice, go sliding down the hill on my ass, feet tangling up... dog barking and howling.... Peckerhead growling and hissing....full charge at my dog....

I go in swinging... slipping... kicking... as I try to get his mini dinosaur off my girl.... I wade in like a warrior... ready to do or die... and Peckerhead is lunging at the dog, I'm body tackling Peckerhead, he's pecking at me, Tara is chasing us both and barking, snapping at the rooster.... A shit show ensues... ice, poop, sloppy water, dog, Peckerhead, and me, the star of the show....

So several bloody fingers after, the dog runs up onto the porch, Peckerhead heads for the hills and I'm lying on the ground wondering what the fuck just happened....

Decisions, Decisions

Well, yesterday was a challenge, I'll tell ya.... I don't know who's gonna go first, Peckerhead or Hubby?

Right now, all bets are off.... Power went out early yesterday morning. A big tree on the edge of property crashed down onto the road, bringing with it a clump of soil the size of a mountain ....so the sound of chainsaws echoed throughout the air at about 5:30... and the sound of a really pissed off rooster... screaming out his morning call... dark... dark... dark... as I made my way to the barn to release and feed the critters.

Flashlight in one hand, grain bucket and hay balanced precariously in the other... fighting with the door, wind about taking me off my feet... branches falling and cracking around me....

Hubby long since gave up and said, screw it, jumped in his truck and headed for work.... Leaving me alone, in the dark, no water, no power and with angry weather and animals... mofo could'a just stayed and babied me but no... he had to do the right thing and go to work... damn him for being a responsible adult!

That's okay, he'll get no supper.... I refuse to even make him a sammich.... He's buying but doesn't know it quite yet... teehheee... yes, I am a brat!

Walking into the barn, remember, dark? Tripped over the food dishes cuz didn't see them in the DARK... and landed on a pile of poop... yeah, it seems the shit follows me... scraping myself off the floor, the goats diving for the spilled grain, trampling me in the process....

Peckerhead gives no warning... and takes a shot at my leg... and I can't see the little flucker amidst the shadows... with only my flashlight... uggh... so I kick out hoping I can dislodge the fluffy little dinosaur and don't get hit in the face with either claw or spur or beak....

So I wrangle my ass up off the floor, my head on a swivel looking for that fluffy little beast... and he's up on his perch, sitting next to Betty, his woman, who is laughing at me.... I shine the flashlight into his eyes and he gives me a menacing growl....

I growl back... well, actually I swear, a long, beautiful string of f-bombs, b-bombs and other un-lady like words... he launches and I bopped him with the flashlight... ready to dance in the dark with this little beast.... The animals are crashing all around me, little goatee horns into my legs, head butts, flapping wings from the two dinosaurs, bleats, snorts, swearing and squawks... a freaking zoo....

So finally I just lose it, completely, embarrassingly, pitifully, lose it, I open the barn door, the hell with the wind, the falling branches, the projectiles flying... he's going out, and I hum him out the door on the end of my boot like a place kicker for the Patriots... and he goes flying!!!!

Then, going to the spigot I plan to fill the water pails... and forget... duh, no power, no running water, so I hump my ass to the little brook and dip the pails... and the handle breaks... in a fit of temper I huck it into the woods... only to have to go fish it out.... I only have one pail... that will soon be remedied.... Hauling water, I finally get the critters settled, fed, watered and poop scooped... covered in shit, I make my way into the house which is filled with smoke... a freakin down draft! My generator quit, my house is filled with smoke, and I'm covered in yuck... no water to shower.... Yes, this day couldn't get any better... or could it?

Oh, yes, yes it does. I haul out the oil lamps, flash lights, candles to prepare for darkness.... And hubby comes home just to trip over the dogs, starts bitching cuz he can't see, the oil lamp does not cast enough light, yadayadayada.... If the shtf... I've decided I'm eating him first... cuz his bitching would set me to the edge of insanity... and in a situation such as that... I'll need my sanity intact....

So, what this power outage has taught me: In the dark, Peckerhead has a definite advantage over this human, I am prepping plenty of bar-b-que sauce because hubby will damn sure get eaten in a shtf situation; I need a new generator and lastly, water is the most precious of all resources.... Oh, and I will survive... but others around me? Well, that is still undecided.

Lessons Learned

There are so many lessons this homestead has taught me... many of them learned hard way....

Lesson One: Never turn your back on a horny buck or a wrangly rooster named Peckerhead.

Lesson Two: There is shit! You will always be covered in shit, in your hair, on your hands, on your boots, your face, pant legs and yes, even up your shirt!

Lesson Three: You are the Queen of the Homestead, but your crown is always, always, always gonna be lop sided and the critters do not respect your title... so jump down off that high horse and dig in!

Okay, so this morning, I was in a hurry. Hubby has an appointment for an inspection for the truck so busily I'm not paying attention and running through my shit patrol duties.

I enter the barn, the babies and mamas are jumping all over me, tripping me up and one mama is crying and bawling like she is in pain... and she is... she wants her buck! The little hoe needs to get laid! I wrestle with them, let Peckerhead and Betty out the door so I don't get attacked as the goats knock me on my ass.... The buck is standing at the door, all bristled up and snorting. I push him out'a the way and bend over to fill the grain dish... he strikes! Jumping on my back and trying to ride me like a fudging pony.... Now, I am five foot two, 115 pounds... and this ain't good! He's a tough little mofo... and buff... strong... I am bitching and swearing, throwing punches and he's riding me like a bronco buster... and I'm getting more pissed off by the minute.... The girls are screaming, Peckerhead is laughing, Betty is watching me with her beady little eyes, squawking... I'm punching the buck, kicking, trying to toss him off of me... hot damn we got a good brew-ha-ha going on and I'm the star of the show!

Finally, I knock his ass back, tackle him to the floor and smack him right in the nose....

"YOU! DO! NOT! RIDE! THE! QUEEN!"

Making my escape I go to the duck pen to feed the Little Nasties... and Peckerhead, decides to slide in behind me and go at the male duck! Feathers are flying, the females are screaming, knives are drawn and blood is spattering.... I dive into the fray, get spurred in the leg, pecked and basically beat to hell with wings and claws and beaks... slipping in shit, got Peckerhead tucked into my arms like a football the male duck flies at my face knocking me into the pool... so I'm in shitty water ass deep, Peckerhead is indignant, I am gagging and have to let him loose to climb out'a the pool. He takes off back at the male duck and they do the street jam.... I'm screaming at them both, hubby is dying with laughter and I've got wet, duck shitty hair.... And damn! I looked so good before feeding time... a little bit of make-up on, a cute shirt, hair in a French braid....

This is the God's of Homesteading reminding me....

The Queen rules the roost but the peons drive the bus! So, off to the shower for the second time today.

Dancing in the Dark

Well ya know when my day starts out stepping in dog poo it's just gonna get worse... it's the way the God of Homesteading reminds me that the poop can splatter the platter at any time. So, Mr. Peckerhead had been minding his manners for the past few days although I thought we'd have a close encounter of the fighting kind yesterday... it didn't happen.... I was tending to my daughters duck... doing surgery... it's got an infected foot so this mama had to put on her nursing skills and do some first aid... yes, on the duck. As my daughter watched on, Peckerhead kept running toward us, circling like a vulture waiting for one of us to fall so he could swoop in and feast on our dead bodies... but, it didn't happen... and I kept shooing him away... and she kept watch... nervously....

This morning though the little fluffy flucker must've had enough of his boring lifestyle and wanted to party a bit.... I walked into the barn after stepping in dog poo, not really paying attention except to the stink coming from my shoe... in my arms I carried the feed pail with the goatee's grain and the chickens pellets, along with a flake of hay... he must've been plotting and planning all night as to how he was gonna murder the prison warden this morning.

He was hiding behind the door, me, unaware of his nefarious intentions, just moseyed on in like it was a normal morning... NOT!

He came out from behind the door... ready... primed... and armed.... No squawk, no growl, no hiss... silent and deadly.... My boot got the first peck... and my leg, the spur.... Buckets got dropped... grain got scattered and I have hay in my hair... as it was tossed up in the air in surprise....

The goatees began ramming my legs, scrabbling for the grain.... Betty White was laughing... and I had my head on a swivel trying to locate that little bastard! You see, I hadn't even turned on the lights yet in the barn and it was still pitch black in there.... I felt like Chuck Norris with a blind fold on... performing feats of karate as I swung at ghosts I couldn't see... but nowhere near as graceful....

I lunged and stepped into the water pail and my boot filled with water as I struggled not to fall... a peck to the back of my leg had me twirling like a drunken pole dancer as I swung around and kicked out at empty air... the flutter of angry wings zipped by my head and I ducked, hitting the floor, smearing poop... chicken and goat... down the front of my shirt.

I grabbed out blindly... screaming in frustration and got a hand full of tail feathers... and I hung on for dear life... bringing him up and over my head as I tossed him against the far wall... and I swear to God I heard him swear at me....

Picking up the bucket from the floor I was armed and ready as he came at me again and this time I hit a home run! I shoved that bucket down over his head and body, squishing him... and stepped on it with my boot so he couldn't escape... and had that little bastard crying UNCLE! Take that you fluffy little asshat! This mama has still got the moves! And my dance is better than his... even with dog poo on my foot!

So we both live to fight another day... he's now in the woods, swearing and threatening me from afar... and me, well I'm just laughing at him, taunting and teasing him to bring it on... yes, I'm feeling a bit rowdy this morning... so I'll give it another go with him gladly this morning....

Death's Door

Well, after sitting my ass in a chair for three days finishing up the rough draft of Valley of Reckoning, this girl needed some good, hard, sweaty outdoor therapy. So, pull a board from the pile, pull the nails, and stack the board. Over, and over and over again. Things were cooking along beautifully, cloudy skies so the sun was baking me, still kinda sticky but not as bad as it has been, my mind is finally quieting down after the grueling three day writing marathon, when out of nowhere I feel a sharp pinch on my leg... and lo and behold, I've got a fluffy flucker attached to me, like a carnivorous leech, stuck, beak buried into my flesh.... No warning, no growl, just Peckerhead staring up at me....

Well, this girl responds before she thinks, especially in pain so I swung my hammer, no, not the claw hammer but rather the big hammer... and knocked that little flucker flying... and guys, for a moment I thought I killed him! He landed with a screech and started kibbying like a dead fish! Shit! Now you all know I have a soft spot for that little beady eyed bastard so I ran over and picked him up thinking I might have to do CPR on the little dude... he hissed, blinked his eyes a couple of times, stared up at me like, WTF! He's still alive, but it was close... old man Death I think was knocking on the door this time....

So now he's hiding from me with Betty White. I'm sure they are plotting my death.

Damn it Peckerhead!

Hot Temper

Well, my temper went from zero to one hundred and ten... in 2.5 seconds flat this morning!

It started off with me opening my eyes at a little after 3 a.m. to a dog that was howling, scratching and yipping at her crate... what in the Hilly Hell! Yelling at her I told her to quiet down... but Oh, no, my dogs listen about as well as a 2 y/o toddler trying to run away when you ask. What have you got in your mouth? So up and out of bed I flew, pissed off and ready to kill... to find out, the little Fuzz butt dog had absolutely nothing fudging wrong with her!

So knowing I wouldn't be able to go back to sleep, I made my first cup of coffee and went out on the dark porch to drink it... and had a fudging June bug, fat and sticky, fly into my hair... Well I danced around like a hooker on crack, arms flailing, coffee cup flying, tripping over my own feet, slamming into the wall trying to dislodge that BIG bug out of my tangled hair....

Screw it! I went back inside, did the heebie jeebie dance for about five minutes while I fixed a second cup of coffee and took a deep breath... finally settling down to the quiet of this early morning....

Well, that fluffy flucker Peckerhead had other ideas....

I hear a ruckus from the coop... an ungodly screeching so I fly out the front door, grab my big stick and launch myself through the coop to only to find him beating up one of my hens.... I fudging lost it! Swing batter batter, swing!

That little bastard flew at me with a murderous glare in his beady reptilian eyes and I swung for the bleachers! Whacked him so hard he tumble rolled across the coop floor... and the hens went screaming out the door with him hot on their heels... so the war was brought to the front yard where Peckerhead, the horny little bastard he is, kept running full tilt toward the hens, jumping on them and getting his groove on... and me batting up and swinging... knocking his fluffy ass Seven ways to Sunday....

He charged at me growling, hissing, spitting... claws aiming for my face... beak at the ready to nail any part of my body he could... me, Eyore bathrobe flying out behind me as I ran making me look like some fudging demented bat as I chased him in a loop around the yard, feet wet from the morning dew on the grass, hens screeching, me using that little flucker as a piñata!

Mind you, I did not have my boots on! So kicking at him with my bare feet, got me nothing but bruised toes... and he was not letting up.... I am trying to protect my hens from this asshole from the fiery depths of rooster hell and he's kicking my ass....

Winded and seeing red I body launched myself at him and pinned him to the ground, my hands around his throat, his claws scrabbling at my stomach... and I punched him....

Yes, I punched my fudging Rooster! And it felt good!

So, now I am able to sit here and drink my first full cup of coffee, my hands are bloody, my back feels like I was kicked by a mule and my Peckerhead is in the barn nursing what I am sure is one hell of a headache.... Yup, some mornings I feel I am just too old for this shit!

Growling

Well, this morning started out per usual except, it wasn't raining. I'm thinking with all this rain, I'm growing fins and gills, mud has now become my best friend. Poop patrol done, critters fed and watered, Peckerhead booted out of the barn... literally... booted, like on the end of my foot, after he charged and hissed and growled at me... not today my nemesis... not today... too much to do.

Started up the fire pit and began burning brush very early, the fog still heavy in the air. Then tackled the project of finishing the floor to my coop.

Hubby: "The floor isn't square."

Me: "Fudge you it isn't square... I measured it a million times."

Hubby: "It's not square."

Me: "Bite me."

He helped me haul the floor frame down to where I wanted it inside the critters pen.

Him: "It's not square."

Me: "I bet you ten dollars it's square."

So he hauls his butt over to get the squary thingy and begins at the corners. Looks up at me with a sheepish grin.

Him: "It's dead on the nuts. Good job."

Me: "Fudge you. I then proceed to place the floor boards down."

Hubby: "Don't forget to look for the crowns."

Now I have no hilly hell idea what this m'fer is talking about... and he's pissing me off... supervisor cap on and telling me what to do? Mmmmmmmm ....

Me: "What the fudge are you talking about?"

Him: "The boards have waves/bend in them... look for the crown and place them on the flatter side."

Me: "Why the fudge didn't you just say that!"

Him: Shaking his head.

So I go to get the 3 inch deck screws.... I'm out. Okay, so I guess I'll hammer and nail the sucker together.

Me: "Where are my nails?"

Him: "I don't know."

Me: "Did you throw my can of nails away?"

You see, last summer I pulled apart a falling down old shed on my property and saved each of the nails I had pulled from the rotted boards... thus the can of MY nails.

Him: "You mean those old rusty nails in the can?"

Me: "Yes! You better had not thrown them away!"

Him: "Relax pack rat... they are in the shed."

So I lay down the boards and begin hammering....

Him: "You can't hit the nail on the head, as I miss the same nail ten times and get my fingers."

Me: "I bet I won't miss your fudging head with this hammer if you keep being a pain in my ass!"

Him: "The floor is not level."

Me: "Bite me."

Him: "It's not level honey."

Me: "OFFS! You are killing me!"

Me: "How can you tell?"

Him: "It doesn't look level."

Me: "Go get the bubble whichamajigy...."

Him: "You mean the level?"

Me: "Fudge you."

He got the level and placed it on the floor... laughing....

Him: "It's nuts dead center."

Me: "I fudging told you."

And this is why we DO NOT work together....

Me: "You had better start running, as I raise the hammer getting ready to throw...."

Him: "Awww honey, don't be like that...."

Me: "Fudge you... you're such a dick sometimes."

Him: Laughing as he walks away....

Spring Love

Well, spring is the season for animal/insect love... and boy Oh, boy.... the peepers are peeping, the mama's are birthing and the horny are hornier if that's at all possible... and Peckerhead is no exception to this spring time ritual or rule.

The morning started out quiet and usual.... Poop patrol, feeding the hungry hordes and just generally working my tail off before the rain we're supposed to get today comes in and forces me to the inside. Once I placed the ladies and the babies in their outdoor pen, I started fire in my fire pit to burn some more brush. Coffee in hand, I settled down to a relaxing few moments by the fire. But, Peckerhead had other ideas... and his idea was to catch a piece of tail from the penned-up hens....

The caterwauling started shortly after I sat down and I went hot footing it to the pen to see what the fudging stir was all about....

Only to see the Peckerhead had hopped the fence and was now having an all-out race to see how much tail he could tag from the four screaming hens.... Now, for those of you who say he's just doing what nature intended... well, kind'a yeah he is... but on this homestead he is not allowed! There are no roses, no sweet whisperings, no foreplay... he is a brute... and when he gets amorous... feathers fly....

My fencing is a hob job of bungy cords and zip ties, and the gate, because of the goats, is bungeed tighter than Fort Knox.... So gaining fast and easy entry is not humanly possible....

I'm at the gate, struggling with bungy cords and screaming at this horny rooster to get the fudge off my girls... my girls are squawking and clucking, flittering and fluttering to get away... and he's hitting them in frenzied record time to get as much tail as he possibly can before I make the launch and tackle his wrangly little ass....

I finally get the gate open, he's running to beat hell to get away from me, I'm tripping over chickens and goats trying to catch him and finally like a linebacker, I launch my whole body in a flying leap and grab his ass feathers... and he turns... ready to give me a go....

Not this time Peckerhead my man.... I roll him under my arm and then huck him like a football over the fence... amidst the applause from some very grateful hens....

As I am stepping out the gate he charges at me and connects with the toe of my boot... growling, hissing, death threats... deadly stares from his beady little eyes....

I stand my ground... guard of the gate, Queen of the Castle, Ruler of the Kingdom.... We circle each other in the timeless dance of life and death... him hurling insults and me daring him to give it his best shot....

Finally, he decides today is not the day to be plucked and thrown in the stew pot... and he makes his merry way back to the back pasture... all the while crowing and prancing his fluffy white ass sassily... and me? Well, I'm guarding the hens... because this horny rooster will try again to gain access over the fence... after all, it tis the season for love. Happy Sunday my friends.

Stuck in The Middle

Here, hold my iced coffee.... Damn it Peckerhead!

Why? Why? God help me why? It is too hot and wayyyyy to early for this... and I'm cranky due to this heatwave... and the cold that's been ripping apart my throat and lungs all week, so it really doesn't surprise me that my fluffy little flucker decides that this week will be the one from hell.

My normal morning routine was shattered... as I hear squawks and screams, profanity and threats from my gnarly rooster. His mate, partner in crime, sexy swanky sidekick, Betty White, has taken to crawling under the barn to lay her eggs. Now I've known she's been doing this and have been meaning to box in the bottom so she cannot gain entry... but just haven't had the energy or time to get around to it. My fault entirely.

Well at the ass crack of dawn that fluffy little asshat decided he was going to wedge his fat body under that barn following Betty White.... Now, this would not have been a problem right? He got in there so he should be able to get out... but noooooooo... he went so deep under that he got wedged by a floor brace... and stuck... he started screaming... and squawking... like he was being killed... and Betty White, laying her eggs, started screaming and squawking... a chorus of mayhem and violence.... I dropped the feed pail and went running thinking something was attacking... grabbed a rake in case I had to go to war defending my little pterodactyl... only to find him stuck... wedged tighter than a cork in a bottle.

Now, I do not like enclosed spaces... nor heights.... nor spiders... especially spiders... but son of a bi**& I was gonna have to get him out'a there... and that meant crawling on my belly under that freakin barn!

So... belly sliding in chicken poop, legs covered with dirt, mosquito's biting at my butt... I am slithering and sliding like a snake to rescue and free Peckerhead....

He is peering at me with those beady little eyes, his neck snaking out pecking at my hands as I reached for him, swearing, threatening, intent even trapped to do me bodily harm.... I am sweating, scraping my back on floor beams, bumping my head, watching for killer spiders.... Betty White is stalking me... circling around my body as she's taking shots at my legs and arms.... Peckerhead is urging her on.... "Kill her, kill her, kill her!" he's chanting....

Finally I grab him by his head and pull, he's kicking and screeching trying to desperately remove a finger or two from my hand....

"I ought to leave you here you little fluffy flucker." I scream as I'm kicking to get Betty off of me and bobbing my head like a boxer to keep away from his beak and claws....

The space is closing in on me and I can feel a

panic attack coming on.... I gasp as the air is running out and I am suffocating.... I have visions of hubby coming home to find only my feet sticking out from under the barn like the wicked witch of the west when the house fell onto her....

Yeah, I know, drama much? But seriously, I think this time I am truly gonna die because of Peckerhead ....and he will have won this war ....I can see the headstone now, "Here lies Nancy, wife, mother and friend, murdered by none other than Peckerhead, her loyal and cantankerous rooster. May she rest in peace."

Well, I did get Peckerhead free, dang near popping his head off in the process... and amidst the scrapes and bumps and bruises, I have lived another day to share with you this tale....

So peace and all that happy crap.... I'm going to take a shower now....

Hugs my friends.

Terrible Toddler

Damn It Peckerhead! WTF! That damn rooster I swear is a reincarnation of a two year old! I am either always running to beat ass away from him or running beat ass to save his fluffy bum.... For the past two weeks he's been hell bent on getting in the duck pen and attacking my Drake... and my females... over the weekend hubby had to get in there and split the two apart and chase him out.... and I've done it several times as well....

Well this afternoon, I'm busy putzing around with projects and I hear a hell'ava screeching and squawking going on... figured something was after my critters... so I hustle down a ladder that is rickety at best because I'm working on my goat addition and take off running....

Peckerhead is in with the ducks, got the Drake pinned, beak full of his feathers... the females are attacking him, it was a freakin pig pile of feathers and squawks.... I dive in, trying to get the girls off of him so I can wrangle his ass off the drake, the girls are biting at my legs, he's hissing and growling and won't fudging let loose, the drake is hissing and fighting back, my legs are getting pummeled (and yes, fudging ducks bite hard)! My ass is on the ground and I'm hugging feathers, his, theirs and just eating shit as I struggle with these little bastards!

Finally I wrap my hand around Peckerheads neck, the drake has a hold of his ass and we're having a tug of war to see who gets him.... Peckerhead nails me with his claws, the duckie girls are still chewing at my legs, I'm punting them like footballs to get them off of me... and now my legs have bruises that once again Hubby will look and say, "Honey how did you get those bruises?" And my reply will be the same, "It wasn't from rough sex with my boyfriend." Yes, he does love my sense of humor....

Thus, Peckerhead is now missing an ass full of feathers... the Drake is pissed off and won't shut up, Betty White is preening her man, and I'm about ready to have a freezer full of critters... uggh.... Ice it baby, walk it off....

The Long Talk

Well, the past two weeks have been a whirl wind of house building, cooking, fighting with barn yard heathens so the stories have been simmering in my head... today started out great. Had plans to do some canning, a ham bone that I needed to boil down for some white bean soup and a bag of dried split peas to make pea soup... the rain came in hard last night so the barnyard is a mixture of soupy poop... and mud... and whatever grossness my animals decide to deposit that the hound hasn't already snorked up... yeah, disgusting.

With the wrangly rascals fed and watered, I tackled the ominous little bastard Peckerhead... he was in a mood and picking the shit out'a Betty White, he even took a swipe at my goatee babies... so I knew he and I needed to have a good old fashioned sit down to hash out his anger issues. Perhaps it is the change of season?

Perhaps the rain? Perhaps just because he's descended from a dinosaur and has a craving for human flesh? Fudged if I know. Regardless, this little asshole with white fluffy wings needed a sharp talking too.

So into the barn I go and he's peering at me with his beady little eyes.

Me: "Look Mister, your behavior is not acceptable. You leave Betty alone or I am going to confine you from your lady!"

Him: Glaring, growling and hissing as he backs into a corner....

Me: "Alright then. We've got that straight."

Him: "The fudge we do!"

He launches for me but guess what? I knew it

was coming! I reared back my leg and went for the scoring punt! And missed. He took a piece of my lower leg... a pecking piece that hurt like a mofo.... So with angry tears I tackled him... and got a horn from one of my goatee babies in the ass... and it hurt... as they frolicked around me like I was their personal spring board....

Coming up off the floor I had Peckerhead by the neck while his claws were reaching for my arms.... "Ha! Sorry sucker! I'm wearing a long sleeve jacket and you can't do shit!" His eyes glared at me as I shook him and we stared at each other face to face, nose to beak.

Me: "I said enough! You will change your nasty attitude!" I yelled ....

Him: "I will fudging kill you human the first chance I get!" (He really did say this!)

Me: "Not if I cook you up in my stew pot first you little jackass!"

Tossing him down he backed into the corner and hissed... hung his head like I'd just neutered him, in shame, in embarrassment, in defeat.... So I'm covered with poop, mud, soaked from the rain and he's subdued for today.... But his eyes promised me, there is always tomorrow for this human to fuck up and become his next victim.

Peckersickle

I think Peckerhead is now a Peckersickle, he's frozen... seriously, didn't even squawk at me this morning... just gave me his little beady eyed glare and said, "Not today bitch, it's too cold…." bwhahahhahahhahhha! Oh, poor little fluffy flucker.

Well, it's been several months since I've written about the homestead. Not because there is nothing going on, but rather, because winter is a closed down season where me and the critters just hunker down and do just the basics in order to stay fed, watered and warm.

The past two days have been bitterly cold, negative temperatures have made both human and animal cranky and a recent snow storm has not improved our mood one iota. Peckerhead, my beloved but nasty little rooster, has for the better part been quiet and low key. That is, until this morning. I don't know if it's the bitter temperatures, the fact that he has been locked tight in the barn to keep his little fluffy ass from getting frostbite or that he was just feeling cantankerous this morning… either way, shit happens and he decided this morning was a shit stirring morning.

I walked into the barn carrying sloshing buckets of water like a drunken sailor due to my vertigo kicking in, my hands already frozen nubs and my face feeling like it was going to crack open from being assaulted by the wind. I was met at the door by four pissy goats and one glaring Peckerhead. I immediately sensed his mood and when he gave my boot a hard peck from his sharp little beak, I knew we'd be in for a tussle.

"Dude… cut it out!" I hissed as he went in for a second peck. I scooted him none to gently with my boot. He growled in response and hopped up on his perch. I went about my duties to the critters, watering, putting out bowls of food and loading in flakes of hay. Mind you, keeping an eyes on the fluffy flucker who was glaring at me as I moved about the barn quietly hissing and growling at me the whole time. Turning, I went to fill his dish and came eye to eye with him.

"Don't even think about it!" I snapped as he fluffed up his feathers and let out another soft growl. Now having had numerous battles with this beast in the past, I knew to back up… and it was a good thing because that little flucker launched himself at my face just as I stepped away from his dish….

With a screech and a swing of my empty feed bucket I knocked that fluffy little flucker across the room like a pro baseball player up to bat… he hit the wall and came at me fighting mad… spurs aimed for my legs, beak aimed for my boots, in a full frontal assault. The goatee girls and little goatee boys stampeded trying to get out of the way of his beak and claws and my kicking legs… me yelling, him hissing and squawking, them scrambling and about taking me off my feet… it was a shit show at its finest.

On the third attack I decided I'd had enough and dove onto that wrangly little white fluff of fury like a linebacker going in for the tackle and grabbed him by his throat. Shaking him I screamed into this glazed eyes.

"I will wreck you dude! It's too fudging cold to be messing with your bad mood this morning! Dear God you will end up in my freezer today!"

With that I tossed him away from me and waiting for retaliation thinking there would be blood… and that little asshat calmly backed into a corner, glared at me while hissing and sat his fat little tush down. So folks, this is how my morning started… fighting with a cantankerous, fluffy fudging rooster who woke up on the wrong side of his perch…. Oh, my.

So now we bring this book to its end, but you

know as well as I do, there is no ending. This saga will continue as I sit back here on my homestead, plotting and planning for future stories to share.

Peckerhead will continue to spend his days tormenting me and challenging me with his rooster ways. And I will continue to stay on the top of my game knowing one day he will knock this queen from her throne.

Peace out, happy homesteading and beware of wrangly roosters.

Made in United States
North Haven, CT
04 November 2026

21772476R00136